Chemometrics

CHEMICAL ANALYSIS

A SERIES OF MONOGRAPHS ON ANALYTICAL CHEMISTRY AND ITS APPLICATIONS

Edited by
J. D. WINEFORDNER

VOLUME 164

Chemometrics
From Basics to Wavelet Transform

FOO-TIM CHAU

Hong Kong Polytechnic University

YI-ZENG LIANG

Central South University

JUNBIN GAO

University of New England

XUE-GUANG SHAO

University of Science and Technology of China

A John Wiley & Sons, Inc., Publication

Published by John Wiley & Sons, Inc., Hoboken, New Jersey.
Published simultaneously in Canada.

For general information on our other products and services please contact our Customer Care Department within the U.S. at 877-762-2974, outside the U.S. at 317-572-3993 or fax 317-572-4002.

Wiley also publishes its books in a variety of electronic formats. Some content that appears in print, however, may not be available in electronic format.

Library of Congress Cataloging-in-Publication Data:

Chemometrics: from basics to wavelet transform/Foo-tim Chau ... [et al.].
  p. cm. — (Chemical analysis; 1075)
Includes bibliographical references.
ISBN 0-471-20242-8 (acid-free paper)
  1. Instrumental analysis—Data processing. 2. Instrumental analysis—Automation.
  3. Wavelets (Mathematics) I. Chau, Foo-tim. II. Chemical analysis; v. 1075.
QD79.I5C44 2004
543′.07′0285—dc21

                2003002429

Printed in the United States of America.
10 9 8 7 6 5 4 3 2 1

CONTENTS

PREFACE

When talking about chemistry, this always leads many to think of doing wet experiments in a laboratory. This was the situation decades ago. Thanks to the development of quantum theory as well as the advancement in electronic and optical devices, chemistry is now evolving into a discipline that corporates both experimentation and modeling together. For instance, nowadays, before synthesizing a new organic compound, database searching can provide information on related reactions to assist in designing viable pathways to synthesize it. In addition, computational chemistry can help determine whether these pathways are favored from the thermodynamic point of view; and QSAR (quantitative structure–activity relationship) studies can help predict the properties of the compound of interest. Similarly, analytical measurements are no longer used only to acquire data from chemical experiments. Signal processing techniques can be used to estimate the precision of these data, extracting more information from the chemical measurements. According to M. Valcarcel, analytical chemistry is a metrological science that develops, optimizes, and applies measuring processes intended to derive both global and partial quality chemical information in order to solve the measuring problems posed.

Chemometrics with the use of statistics and related mathematical techniques forms a new area in chemistry. According to D. L. Massart, its targets are to design or select optimal measurement procedures and experiments as well as to extract a maximum of information from chemical data. With these unique features and applications, some believe that chemometrics provides an important theoretical background for analytical chemistry.

In recent years, wavelet transform (WT), a new mathematical technique, has been widely used in engineering sciences owing to its localization properties in both the frequency and time domains. It was introduced to chemistry in 1990s and has now attracted the attention of many chemists. Prior to January 2003, over 370 chemistry papers and references related to WT had been published.

Many chemists facing sophisticated practical problems are unfamiliar with the chemometric methods, especially using such new approaches as WT, available for solving their problems. They would be happy if they could find the appropriate methods they needed, but where to find them? This seems to be one of the major obstacles in the way of wide applications of chemometric methods in chemistry. The famous Chinese philosopher Guo-wei Wang (1877–1927) cited the lyrics of Song Dynasty to describe different extent of a scholar's learning. According to Wang, the highest extent of knowledge is that described by Qi-ji Xin (1140–1207); from the tune "Green Jade Cup," Lantern Festival:

> But in the crowd once and again
>> I look for her in vain.
> When all at once I turn my head,
>> I find her there where lantern light is dimly shed.

It is not so easy for the average chemist to reach such an extent of learning in the mathematical background of chemometric theory. Fortunately, this book on chemometrics from the basics to wavelet transform by Professor F. T. Chau's team is written in a tutorial manner with many examples provided to clarify the theory and methods described. The basic theory of WT and its applications to analytical chemistry are described. In addition, the fundamentals of chemometrics and various common signal processing techniques are provided to help readers learn more about the applications of this new mathematical technique. The basic fundamentals of vector and matrix operations and the mathematical programming language MATLAB are also provided in the Appendix to enable newcomers to the field to derive more from the contents of this book. In addition, computer codes are provided for some topics to help the readers to see how the proposed algorithms work in real life. Relevant literature references are also listed at the end of each chapter.

It is really a great honor for me to be invited to write these lines for the book. The authors have undertaken the large task of surveying the subject to provide a valuable reference book for chemists, biochemists, and postgraduate students. In fact, even the most modern innovations of WT have found a place in this concise volume. With its own distinctiveness, this book is indeed a very welcome addition to the existing literature on chemometrics.

Professor of Chemistry   Ru-Qin Yu
Member of Chinese Academy of Sciences
Hunan University
Changsha, People's Republic of China

CHAPTER

1

INTRODUCTION

1.1. MODERN ANALYTICAL CHEMISTRY

1.1.1. Developments in Modern Chemistry

The field of chemistry is currently facing major changes. As we know, optical, mechanical, and microelectronic technologies have advanced rapidly in recent years. Computer power has increased dramatically as well. All these developments, together with other factors, provide a new opportunity but also challenge to chemists in research and development.

A recent (as of 2003) development in the pharmaceutical industry is the use of combinatorial synthesis to generate a library of many compounds with structural diversity. These compounds are then subjected to high-throughput screening for bioassays. In such a process, tremendous amounts of data on the structure–activity relationship are generated. For analytical measurements, a new, advanced, modern technology called *hyphenated instrumentation* using two or more devices simultaneously for quantitative measurement has been introduced [1]. Examples of this technique are the high-performance liquid chromatography–diode array detector system (HPLC-DAD), gas chromatography with mass spectrometry (GC-MS), and liquid chromatography coupled with mass spectrometry such as LC-MS and LC-MS-MS. Huge amounts of data are generated from these pieces of equipment. For example, the Hewlett-Packard (HP) HPLC 1100 instrument with a diode array detector (DAD) system (Agilent Technology Inc., CA) produces 1.26 million spectrochromatographic data in a 30-min experimental run with a sampling rate of 5 Hz, and a spectral range of 190–400 nm with a resolution of 1 data item per 2 nm. To mine valuable information from these data, different mathematical techniques have been developed. Up to now, research and development of this kind with the application of statistical and mathematical techniques in chemistry has been confined mainly to analytical studies. Thus, our

Chemometrics: From Basics To Wavelet Transform, edited by Foo-tim Chau, Yi-zeng Liang, Junbin Gao, and Xue-guang Shao. Chemical Analysis Series, Vol. 164. ISBN 0-471-20242-8. Copyright © 2004 John Wiley & Sons, Inc.

discussion will focus on analytical chemistry but other disciplines of chemistry will also be included if appropriate. The main content of this book provides basic chemometric techniques for processing and interpretation of chemical data as well as chemical applications of advanced techniques, including wavelet transformation (WT) and mathematical techniques for manipulating higher-dimensional data.

1.1.2. Modern Analytical Chemistry

Modern analytical chemistry has long been recognized mainly as a measurement science. In its development, there are two fundamental aspects:

1. From the instrumental and experimental point of view, analytical chemistry makes use of the basic properties such as optics, electricity, magnetism, and acoustic to acquire the data needed.

   In addition,

2. New methodologies developed in mathematical, computer, and biological sciences as well as other fields are also employed to provide in-depth and broad-range analyses.

Previously the main problem confronting analytical scientists was how to obtain data. At that time, measurements were labor-intensive, tedious, time-consuming, and expensive, with low-sensitivity, and manual recording. There were also problems of preparing adequate materials, lack of proper techniques, as well as inefficient equipment and technical support. Workers had to handle many unpleasant routine tasks to get only a few numbers. They also had to attempt to extract as much information as possible about the structure, composition, and other properties of the system under investigation, which was an insurmountable task in many cases. Now, many modern chemical instruments are equipped with advanced optical, mechanical, and electronic components to produce high-sensitivity, high-quality signals, and many of these components are found in computers for controlling different devices, managing system operation, data acquisition, signal processing, data interpretation in the first aspect and reporting analytical results. Thus the workload on analytical measurement mentioned above (item 1 in list) is reduced to minimum compared to the workload typical decades ago.

After an analytical measurement, the data collected are often treated by different signal processing techniques as mentioned earlier. The aim is to obtain higher quality or "true" data and to extract maximum amount of meaningful information, although this is not easy to accomplish. For instance, in an HPLC study, two experimental runs were carried out on the same sample mixture. The two chromatograms acquired usually differed from each other to a certain extent because of the variations in instrumentation, experimental conditions, and other factors. To obtain quality results that are free from these disturbances, it is a common practice to carry out data preprocessing first. The techniques involved include denoising, data smoothing, and/or adjustment of baseline, drift, offset, and other properties. Methods such as differentiation may then be applied to determine more accurate retention times of peaks, especially the overlapping peaks that arise from different component mixtures. In this way, some of these components may be identified via their retention times with a higher level of confidence through comparison with those of the standards or known compounds. If the peak heights or peak areas are available, the concentrations of these components can also be determined if the relevant calibration curves are available. Statistical methods can also help in evaluating the results deduced and to calculate the level of confidence or concentrations of the components being identified. All these data obtained are very important in preparing a reliable report for an analytical test. Data treatment and data interpretation on, for instance, the HPLC chromatograms as mentioned above form part of an interdisciplinary area known as *chemometrics*.

1.1.3. Multidimensional Dataset

Many analytical instruments generate one-dimensional (1D) data. Very often, even if they can produce multidimensional signals, 1D datasets are still selected for data treatment and interpretation because it is easier and less time-consuming to manipulate them. Also, most investigators are used to handle 1D data. Yet, valuable information may be lost in this approach.

Figure 1.1 shows the spectrochromatogram obtained in a study of the herb Danggui (*Radix angeliciae sinensis*) [2] by using the Hewlett-Packard (HP) HPLC-DAD model 1100 instrument. Methanol was utilized for sample extraction. In carrying out the experiment, a Sep-Pak C_{18} column was used and the runtime was 90 min. The two-dimensional (2D) spectrochromatogram shown in Figure 1.1 contains 2.862 million data points [3]. It looks very complicated and cannot be interpreted easily just by visual inspection. As mentioned earlier, many workers simplify the job by selecting a good or an acceptable 1D chromatogram(s) from Fig. 1.1 for analysis. Figure 1.2

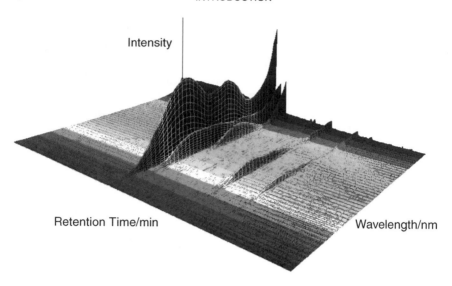

Figure 1.1. The 2D HPLC chromatogram of Danggui.

shows the 1D chromatograms selected with the measured wavelengths of 225, 280, and 320 nm, respectively. However, which one should be chosen as the fingerprint of Danggui is not an easy question to answer since the profiles look very different from one another. The variation in these chromatographic profiles is due mainly to different extents of ultraviolet absorption of the components within the herb at different wavelengths. From an information analysis [3], Figure 1.2b is found to be the best chromatogram. Yet, the two other chromatograms may be useful in certain aspects.

Methods for processing 1D data have been developed and applied by chemists for a long time. As previously mentioned, noise removal, background correction, differentiation, data smoothing and filtering, and calibration are examples of this type of data processing. Chemometrics is considered to be the discipline that does this kind of job. With the growing popularity of hyphenated instruments, chemometric methods for manipulating 2D data have been developing. The increasing computing power and memory capacity of the current computer further expedites the process. The major aim is to extract more useful information from mountainous 2D data. In the following section, the basic fundamentals of chemometrics are briefly introduced. More details will be provided in the following chapters.

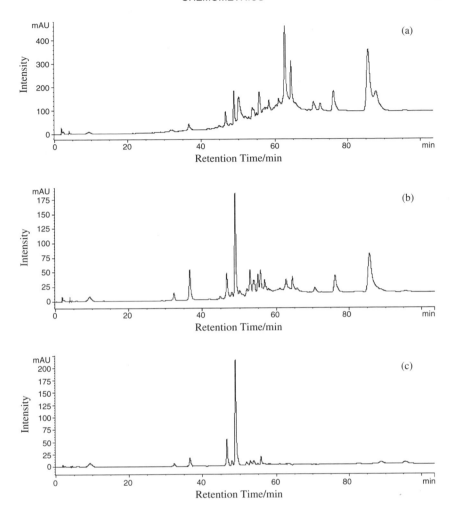

Figure 1.2. The HPLC chromatogram of Danggui measured at (a) 225 nm, (b) 280 nm, and (c) 320 nm.

1.2. CHEMOMETRICS

1.2.1. Introduction to Chemometrics

The term *chemometrics* was introduced by Svante Wold [4] and Bruce R. Kowalski in the early 1970s [4]. Terms like *biometrics* and *econometrics* were also introduced into the fields of biological science and economics. Afterward, the International Chemometrics Society was established. Since then, chemometrics has been developing and is now widely applied to different fields of chemistry, especially analytical chemistry in view of the

numbers of papers published, conferences and workshops being organized, and related activities. "A reasonable definition of chemometrics
remains as how do we get chemical relevant information out of measured
chemical data, how do we represent and display this information, and how
do we get such information into data?" as mentioned by Wold [4]. Chemometrics is considered by some chemists to be a subdiscipline that provides
the basic theory and methodology for modern analytical chemistry. Yet, the
chemometricans themselves consider chemometrics is a new discipline of
chemistry [4]. Both the academic and industrial sectors have benefited
greatly in employing this new tool in different areas.

The Howery and Hirsch [5] in the early 1980s classified the development
of the chemometrics discipline into different stages. The first stage is
before 1970. A number of mathematical methodologies were developed
and standardized in different fields of mathematics, behavioral science, and
engineering sciences. In this period, chemists limited themselves mainly to
data analysis, including computation of statistical parameters such as the
mean, standard deviation, and level of confidence. Howery and Hirsch, in
particular, appreciated the research on correlating vast amounts of chemical data to relevant molecular properties. These pioneering works form the
basis of an important area of the quantitative structure–activity relationship
(QSAR) developed more recently.

The second stage of chemometrics falls in the 1970s, when the term
chemometrics was coined. This new discipline of chemistry (or subdiscipline of analytical chemistry by some) caught the attention of chemists,
especially analytical chemists, who not only applied the methods available for data analysis but also developed new methodologies to meet
their needs. There are two main reasons why chemometrics developed
so rapidly at that time: (1) large piles of data not available before could be
acquired from advanced chemical instruments (for the first time, chemists
faced bottlenecks similar to those encountered by social scientists or
economists years before on how to obtain useful information from these
large amounts of data) and (2) advancements in microelectronics technology within that period. The abilities of chemists in signal processing and
data interpretation were enhanced with the increasing computer power.

The future evolution of chemometrics was also predicted by Howery
and Hirsch in their article [5] and later by Brown [4]. Starting from the
early 1980s, chemometrics were amalgamated into chemistry courses
for graduates and postgraduates in American and European universities. In addition, it became a common tool to chemists. Since the early
1980s, development of the discipline of chemometrics verified the original predictions. Chemometrics has become a mainstay of chemistry
in many universities of America and Europe and some in China and

other countries. Workshops and courses related to chemometrics are held regularly at conferences such as the National Meetings of American Chemical Society (ACS) and the Gordon Conferences, as well as at symposia and meetings of the Royal Society of Chemistry and International Chemometrics Society. For instance, four courses were offered under the title "Statistics/Experimental Design/Chemometrics" in the 226th ACS National Meeting held in New York in September 2003 [*http://www.acs.org*]. The course titles are "Chemometric Techniques for Qualitative Analysis," "Experimental Design for Combinatorial and High-Througput Materials Development," "Experimental Design for Productivity and Quality in R&D," and "Statistical Analysis of Laboratory Data." Furthermore, chemometrics training courses are held regularly by software companies like such as CAMO [6] and PRS [7]. In a review article [8] on the 25 most frequently cited books in analytical chemistry (1980–1999), four are related to chemometrics: *Factor Analysis in Chemistry* by Malinowski [9], *Data Reduction and Error Analysis for the Physical Sciences* by Bevington and Robinson [10], *Applied Regression Analysis* by Draper and Smith [11], and *Multivariate Calibration* by Martens and Naes [12] with rankings of 4, 5, 7, and 16, respectively. The textbook *Chemometrics: Statistics and Computer Applications in Analytical Chemistry* [13] by Otto was the second most popular "bestseller" on analytical chemistry according to the Internet source *www.amazon.com* on February 16, 2001. The Internet source *www.chemistry.co.nz* listed "Statistics for Analytical Chemistry" by J. Miller and J. Miller as one of the eight analytical chemistry bestsellers on January 21, 2002 and February 10, 2003.

Chemometricians have applied the well-known approaches of multivariate calibration, chemical resolution, and pattern recognition for analytical studies. Tools such as partial least squares (PLS) [14], soft independent modeling of class analogy (SIMCA) [15], and methods based on factor analysis, including principal-component regression (PCR) [16], target factor analysis (TFA) [17], evolving factor analysis (EFA) [18,19], rank annihilation factor analysis (RAFA) [20,21], window factor analysis (WFA) [22,23], and heuristic evolving latent projection (HELP) [24,25] have been introduced. In providing the basic theory and methodology for analytical study, its evolution falls into main two categories: (1) development of new theories and algorithms for manipulating chemical data and (2) new applications of the chemometrics techniques to different disciplines of chemistry such as environmental chemistry, food chemistry, agricultural chemistry, medicinal chemistry, and chemical engineering. The advancements in computer and information science, statistics, and applied mathematics have introduced new elements into chemometrics. Neural networking [26,27], a mathematical technique that simulates the transmission of signals within the biological system; genetic algorithm [28–30]; and simulated anneal-ing [31,32] are examples of this methodology. Wavelet analysis [33] of information science, a robust method of statistics and image analysis of engineering and medical sciences, is another example.

Common methods for processing 1D data are reviewed and discussed in Chapter 2 to provide the readers with the basic knowledge of chemomet-rics. Chapter 3 discusses an area of current interest in signal processing involving generation of 2D data using modern analytical instruments. The basic theory and applications of WT in chemistry are introduced in Chapters 4 and 5, respectively. WT-related papers are briefly reviewed in Chapter 5 to give the readers an overview of WT applications in chem-istry. Examples of various chemometric methods are provided with lists of MATLAB codes [34] for the calculations involved.

1.2.2. Instrumental Response and Data Processing

Instrumental response in analytical study usually necessitates some form of mathematical manipulation to make the manipulations more mean-ingful to chemists. For instance, the absorbance of a peak deduced from its ultraviolet–visible spectrum can be transformed into a chemical variable of concentration of the related compound. Similarly, the HPLC peak height or area of a component can be correlated to its concentra-tion. Considering the latter case, if the HPLC peak of a one-component system is free from overlapping and other interference, then the concen-tration of the component can be determined when the relevant calibration curve is available. In this case, only one concentration or chemical vari-able is involved and the system is regarded as a *univariate* one by chemometricians.

Figure 1.3 shows the chromatogram of the ginsenoside extract of an American Ginseng sample as acquired by the HP HPLC model 1100 instrument. It consists mainly of several peaks corresponding to the gin-senosides of Re, Rg1, Rf, Rb1, Rc, Rb2, and Rd. The peaks with retention times greater than 15 min are well separated from one another and are free from impurities. Hence, each one can be considered as a one-component system that can be treated using univariate methods under an ideal situation. Yet, the profile around 6 min shows two overlapping peaks that should be classified as a multivariate system in data analy-sis. Usually, analytical chemists try their best to adjust the experimental conditions to separate these overlapping peaks. Yet, it is time-consuming and not easy to achieve this, especially for complicated systems such as herbal medicine, which has many constituents. Thus, different multivariate

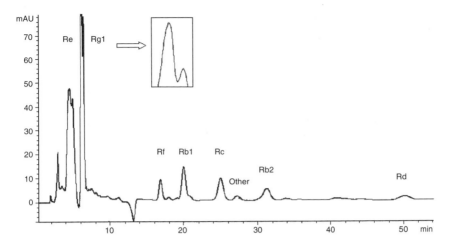

Figure 1.3. The HPLC chromatogram of the ginsenoside extract of an American Ginseng sample.

analysis techniques have been developed to accomplish the target. The powerful computers currently available considerably help in this respect. The multivariate approach is currently considered as one of the developing areas of chemometrics for tackling problems for complicated chemical systems such as in the example cited above.

Most data obtained from analytical measurements are now represented in digital rather than analog form as previously. Thus, only manipulation of digital data is considered throughout this book.

1.2.3. White, Black, and Gray Systems

Mixture samples commonly encountered in analytical chemistry fall into three categories, which are known collectively as the white–gray–black multicomponent system [35] according to the information available. The samples labeled "white" have the following characteristics. All spectra of the component chemical species in the sample, as well as the impurities or interferents present, are available. Concentrations of all the selected analytes are also known for the investigation. In this system, the major aim is to quantitatively determine the concentrations of some or all the chemical components of the sample mixture. Chemometric methods for this kind of analysis are relatively mature.

The system with no a priori information regarding the chemical composition is classified as a "black" one. The aim of the chemometric treatment could be to determine simultaneously spectra (resolution) and concentrations (quantification) of all the chemical components. To most, it is impossible to achieve this. The advantages of having hyphenated instrumentation data as mentioned previously is that one can decide the number of analytes in peak clusters and, in many cases, resolve them into single components without access to standards if appropriate chemometric techniques are used.

The "gray" analytical system covers samples between the two types def ined above. No complete knowledge is available for the chemical composition or spectral information. The aim of signal processing in this system is to determine quantitatively the selected analytes in the presence of unknown interferents or components. In this way, spectra or models for the analytes are available. Yet, no information regarding possible interferents or other unknown chemicals in the samples is on hand.

1.3. CHEMOMETRICS-BASED SIGNAL PROCESSING TECHNIQUES

1.3.1. Common Methods for Processing Chemical Data

Chemical signals obtained from analytical measurements are usually recorded as chromatograms, spectra, kinetic curves, titration curves, or in other formats with the domain of time, wavelength, or frequency. Signal processing may be carried out in the "online" or "offline" mode. For the online mode, the data acquired are treated immediately and the time spent has to be relatively short for monitoring or controlling the system in real time. As for the offline mode, a complete set of data is collected from measurements and manipulation of data is carried out afterward.

In general, quality of the raw experimental data is evaluated with the use of statistical analysis. Parameters such as mean, variance, and standard deviation are utilized very often for this purpose. The assessment can be further carried out by some methods such as Student's t-test and F-test, where the data are assumed to have the Gaussian distribution. With availability of the full dataset, signal processing techniques such as smoothing and filtering methods, transformation methods, and numerical treatment methods can further be used to analyze and enhance the data. The aims are to improve the signal-to-noise ratio (SNR), to convert the data into more physically meaningful form, to extract useful and/or accurate information, to classify objects, and so on. In Chapter 2, signal processing of this type on 1D data will be discussed in detail. As for 2D data, techniques for background correction, congruence analysis, least-square fitting, differentiation, and resolution of overlapping signal will be described.

1.3.2. Wavelets in Chemistry

It is well known that Fourier transform (FT) has played an important role in chemical data processing. FT converts a signal from one form into another form and simplifies the complex signal for chemists. For example, FT-IR and FT-NMR spectra are obtained by transforming the signals measured in the time domain into the frequency domain. Furthermore, development of several signal processing methods such as filtering, convolution/deconvolution, and derivative calculation was also based on Fourier transform.

A *wavelet* is defined as a family of functions derived from a basic function, called the *wavelet basis function*, by dilation and translation. Wavelet basis functions are those functions with some special properties such as orthogonality, compact support, symmetry, and smoothness. Wavelet transform (WT) [36–39] is a projection operation of a signal onto the wavelet. In some respects, WT is simply an analog to FT. The only difference is the basis functions. In FT, the trigonometric (sine and cosine) functions are the basis functions, while the basis function in WT is the wavelet. Therefore, a large number of basis functions are available as compared with FT.

The most outstanding characteristic of the WT is the localization property in both time and frequency domains, while FT is localized only in the frequency domain. With proper identification, WT may be used to zoom in or zoom out a signal at any frequency and in any small part. Thus, it can be called a *mathematical microscope* of analytical signals. A complex signal can be decomposed into its components with different frequencies by using the WT technique. This provides us an opportunity to examine any part of the signal of interest. Most of the WT applications on processing chemical signals are based on the dual-localization property. Generally, the signal is composed of baseline, noise, and chemical signals. Since the frequencies of the three parts as mentioned above are significantly different from one another, it is not difficult to use the WT technique by removing the high-frequency part from the signal for denoising and smoothing and by removing the low-frequency part for baseline correction. Similarly, one can also use WT to extract the high-resolution information from a low-resolution signal by eliminating both the high-frequency noise and the low-frequency background.

Another property of the WT is that it can be turned into *sparse expansions*, which means that any signal can be quite accurately represented by a small part of the coefficients derived. This property makes WT an effective tool for data compression. Several applications of WT, such as derivative calculation by using the discrete wavelet transform (DWT) and continuous wavelet transform (CWT), have also been developed based on the special properties of the wavelet basis functions. Furthermore, the combined methods of WT and other chemometric methods were reported. For example, the wavelet neural network (WNN) provides a new network model for data compression, pattern recognition, and quantitative prediction. Taking advantage of the WT in data compression, the efficiency of the artificial neural network (ANN), partial least squares (PLS), chemical factor analysis (CFA), and other procedures can be improved by combining these methods with WT. A wavelet toolbox is available from MathWorks, Inc. [33] to carry out WT calculations.

WT became a hot issue in chemistry field during the late 1980s. More than 370 papers and several reference books were published from 1989 to 2002. In these published works, both the theory and the applications of WT were introduced. WT was employed mainly for signal processing in different fields of analytical chemistry, including flow injection analysis (FIA), high-performance liquid chromatography (HPLC), capillary electrophoresis (CE), ultraviolet–visible spectrometry (UV–vis), infrared spectrometry (IR), Raman spectroscopy, photoacoustic spectroscopy (PAS), mass spectrometry (MS), nuclear magnetic resonance (NMR) spectrometry, atomic absorption/emission spectroscopy (AAS/AES), X-ray diffraction/spectroscopy, potentiometric titration, voltammetric analysis, and analytical image processing. It has also been employed to solve certain problems in quantum chemistry and chemical physics. More references and online resources can be found in Chapter 5.

1.4. RESOURCES AVAILABLE ON CHEMOMETRICS AND WAVELET TRANSFORM

1.4.1. Books

Tutorial and Introductory

Beebe, K., R. Pell, and M. B. Seasholtz, *Chemometrics, A Practical Guide*, Wiley, New York, 1998.

Brereton, R., *Data Analysis for the Laboratory and Chemical Plant*, Wiley, New York, 2003.

Bevington, P. R., and D. K. Robinson, *Data Reduction and Error Analysis for the Physical Sciences*, 3rd ed., McGraw-Hill, New York, 2002.

Davies, L., *Efficiency in Research, Development and Production: The Statistical Design and Analysis of Chemical Experiments*, Royal Society of Chemistry, London, 1993.

Draper, N., and H. Smith, *Applied Regression Analysis*, 3rd ed., Wiley, New York, 1999.

Jackson, J. E., *A User's Guide to Principal Components*, Wiley, New York, 1991.

Kramer, R., *Chemometric Techniques for Quantitative Analysis*, Wiley, New York, 1998.

Malinowski, E., *Factor Analysis in Chemistry*, 3rd ed., Wiley, New York, 2002.

Meloun, M., J. Militky, and M. Forina, *Chemometrics for Analytical Chemistry*, Vols. 1 and 2, *PC-Aided Statistical Data Analysis*, Ellis Horwood, New York, 1992, 1994.

Miller, J. N., and J. C. Miller, *Statistics and Chemometrics for Analytical Chemistry*, 4th ed., Prentice-Hall, London, 2000.

Otto, M., *Chemometrics: Statistics and Computer Application in Analytical Chemistry*, Wiley-VCH, Weinheim, 1999.

Sharaf, M. A., D. L. Illman, and B. R. Kowalski, *Chemometrics*, Chemical Analysis Series Vol. 82, Wiley, New York, 1986.

Vandeginste, B. G. M., D. L. Massart, L. M. C. Buydens, S. De Jong, P. J. Lewi, and J. Smeyers-Verbeke, *Handbook of Chemometrics and Qualimetrics*, Parts A and B, Elsevier, Amsterdam, 1998.

Walnut, D. F., *An Introduction to Wavelet Analysis*, Birkhuser, Boston, 2002.

Advanced

Deming, S. N., and S. L. Morgan, *Experimental Design: A Chemometrics Approach*, 2nd ed., Elsevier, Amsterdam, 1993.

Goupy, J. L., *Methods for Experimental Design: Principles and Applications for Physicists and Chemists*, Elsevier, Amsterdam, 1993.

Jurs, P. C., ed., *Computer-Enhanced Analytical Spectroscopy*, Vol. 3, Plenum Press, New York, 1992.

Liang, Y. Z., *White, Grey and Black Multicomponent System and Their Chemometric Algorithms* (in Chinese), Hunan Publishing House of Science and Technology, Changsha, 1996.

Liang, Y. Z., R. Nortvedt, O. M. Kvalheim, H. L. Shen, and Q. S. Xu, eds., *New Trends in Chemometrics*, Hunan University Press, Changsha, 1997.

Martens, H., and M. Martens, *Multivariate Analysis of Quality: An Introduction*, Wiley, New York, 2001.

Meuzelaar, H. L. C., and T. L. Isenhour, eds., *Computer-Enhanced Analytical Spectroscopy*, Plenum Press, New York, 1987.

Meuzelaar, H. L. C., ed., *Computer-Enhanced Analytical Spectroscopy*, Vol. 2, Plenum Press, New York, 1990.

Rouvray, D. H., ed., *Fuzzy Logic in Chemistry*, Academic Press, San Diego, 1997.

Schalkoff, R., *Pattern Recognition Statistical Structural and Neural Approaches*, Wiley, New York, 1992.

Wakzak, B., ed., *Wavelets in Chemistry*, Elsevier, Amsterdam, 2000.

Wilkins, C. L., ed., *Computer-Enhanced Analytical Spectroscopy*, Vol. 4, Plenum Press, New York, 1993.

General References

Chatfield, C., and A. J. Collins, *Introduction to Multivariate Analysis*, Chapman & Hall, London, 1989.

DeMuth, J. E., *Basic Statistics and Pharmaceutical Statistical Application*, Wiley, New York, 1999.

Spiegel, M. R., *Schaum's Outline of Theory and Problems of Statistics*, McGraw-Hill, New York, 1999.

Related Topics

Adams, M. J., *Chemometrics in Analytical Spectroscopy*, Royal Society of Chemistry, London, 1995.

Ciurczak, E. W., and J. K. Drennen, *Pharmaceutical and Medical Applications of NIR Spectroscopy*, Marcel Dekker, New York, 2002.

Einax, J. W., H. W. Zwanziger, and S. Geiss, *Chemometrics in Environmental Analysis*, VCH, Weinheim, 1997.

Mark, H., *Principles and Practice of Spectroscopic Calibration*, Wiley, New York, 1991.

Zupan, J., and J. Gasteiger, *Neural Networks for Chemists: An Introduction*, VCH, Weinheim, 1993.

1.4.2. Online Resources

General Websites

http://www.spectroscopynow.com/Spy/basehtml/SpyH/1,2142,2-0-0-0-0-home-0-0,00.html. *Chemometrics World* provides comprehensive Web resources, including online traning, job opportunities, software links, conferences, and discussion forums. Under the "Education" menu, there is a glossary list compiled by Bryan Prazen as well as tutorials.

http://www.acc.umu.se/~tnkjtg/chemometrics/. The *Chemometrics* homepage is a portal site edited by Johan Trygg. It provides links to useful Webpages related to the subject as well as software links. It also maintains a list for upcoming conferences.

http://www.chemometrics.com/. This is the Website of *Applied Chemometrics*. Information on good introductory books, software links, training courses, and other sources is provided.

http://iris4.chem.ohiou.edu/. This is the official Website of the North American chapter of the International Chemometrics Society (NAmICS).

http://www.amara.com/current/wavelet.html. Amara's "Wavelet" page provides comprehensive online information on wavelets. The "Beginners Bibliography" is a good place to start if you are not familiar with signal processing. The software list is extensive.

Journals

http://www.interscience.wiley.com/jpages/0886-9383/. This is the *Journal of Chemometrics*'s homepage. A full-text electronic version is available to subscribers for articles since 1997.

http://www.elsevier.nl/inca/publications/store/5/0/2/6/8/2/. This is the *Journal of Chemometrics and Intelligent Laboratory*'s homepage. All articles (including those in Vol. 1, published in 1986) are available in Acrobat format for subscribers to download.

http://pubs.acs.org/journals/jcisd8/index.html. This is the homepage of the *Journal of Chemical Information and Computer Sciences*. All articles (including Vol. 1, published in 1975) are available in Acrobat format for subscribers to download.

http://www.wavelet.org/. The *Wavelet Digest* is an online journal devoted to the wavelet community. It has preprints and announcements of new books, software, jobs, and conferences.

Tutorials and Personal Homepages

http://www.galactic.com/. Thermo Galactic markets a comprehensive data processing management software package, GRAMS/AI, which can work with data files from multiple techniques, including UV, UV–vis, IR, FT-IR, NIR, NMR, LC, GC, HPLC, DAD, CE, and mass and Raman spectrometry. There is a Chemometrics extension available called PLSplus/IQ. This Website provides descriptions of the algorithms used in the programs (*http://www.galactic.com/Algorithms/default.asp*), which is also a good introduction to these topics.

http://www.chm.bris.ac.uk/org/chemometrics/pubs/index.html. This is the Website of the Bristol Chemometrics research groups. An evaluation version of their Excel add-in software is available for download with some exercises provided.

http://ourworld.compuserve.com/homepages/Catbar/int_chem.htm. This Website publishes a report entitled "An introduction to chemometrics," by Brian A. Rock in October 1985.

http://www.chem.duke.edu/~clochmul/tutor1/factucmp.html. This Webpage provides introductory notes on factor analysis by Charles E. Reese and C. H. Lochmüller.

http://ull.chemistry.uakron.edu/chemometrics/. This is lecture presentations of a chemometrics course by James K. Hardy of the University of Akron.

http://www.models.kvl.dk/users/rasmus/. This is the personal Website of Professor Rasmus Bro. He offers a free monograph *Multi-way Analysis in the Food Industry*, which describes several applications of wavelets in the food industry.

http://www.ecs.syr.edu/faculty/lewalle/tutor/tutor.html. This is a tutorial on continuous wavelet analysis of experimental data; plots are available on most of the topics illustrating wavelet techniques in simplified format, without unnecessary mathematical sophistication.

http://rcmcm.polyu.edu.hk/wavelet.html. The *Wavelet Transform in Chemistry* provides a comprehensive literature survey of the application of wavelet transform in different area of chemistry.

1.4.3. Mathematics Software

A brief introduction to some useful software packages for chemometrics techniques and wavelet transform is given below. Some of the packages are even freely available for public download. Many of them have specialized wavelet extensions available. Thus, we hope that readers can find the appropriate tools to test the data processing techniques discussed in this book. We would recommend MATLAB if readers can access the software; however, other packages can perform the calculations equally well.

MATLAB (*http://www.mathworks.com/*). MATLAB is a popular and powerful tool in the field marketed by Mathworks. It integrates mathematical computing, visualization, and a powerful language to provide a flexible computing environment. This makes MATLAB an extremely useful tool for exploring data and creating algorithms. The open architecture makes MATLAB extensible, and many "toolboxes" have been developed by MATLAB users. All the examples provided in this book are also implemented in MATLAB script.

There are many toolboxes available for applying chemometrics techniques. A detailed list can be found in *http://www.mathtools.net/MATLAB/Chemometrics/index.html*. Examples including the "chemometrics" toolbox, the "factor analysis" (FA) toolbox, and the "principal least-squares" toolbox provide functions for multivariate analysis such as principal-component analysis (PCA), principal-component regression (PCR), and partial least squares (PLS).

Mathworks also markets a toolbox for wavelet analysis. The "wavelet" toolbox is a full-featured MATLAB [graphical user interface (GUI) and command-line] toolbox originally developed by Michel Misiti, Yves Misiti, Georges Oppenheim, and Jean-Michel Poggi. The toolbox provides continuous wavelet transforms (CWTs), discrete wavelet transforms (DWTs), user-extendible selection of wavelet basis functions, wavelet packet transforms, entropy-based wavelet packet tree pruning for "best tree" and "best level" analysis, and soft and hard thresholding Denoising (please refer to Chapter 5). One- and two-dimensional wavelet transforms are supported so that they can be used for both signal and image analysis and compression.

Apart from the "official" wavelet toolbox from Mathworks, there are also many other wavelet toolboxes created by different researchers. Examples in Chapter 5 are built on one of them, the WaveLab developed by researchers of Stanford University. The current version of WaveLab is at level 8.02, yet the examples in Chapter 5 are based on an earlier version. The toolbox is available from the developer's Website, *http://www-stat.stanford.edu/~wavelab/*. Many other toolboxes are available, including the Rice Wavelet Toolbox (*http://www-dsp.rice.edu/software*), the Wavekit by Harri Ojanen (*http://www.math. rutgers.edu/~ojanen/wavekit/*), and the Mulitwavelet package by Vasily Strela (*http://www.mcs.drexel.edu/~vstrela/*).

Scilab (*http://www-rocq.inria.fr/scilab/*). Scilab is a software package for numerical computations providing a powerful open computing environment for engineering and scientific applications similar to MATLAB. Yet, it is distributed freely with binary versions for UNIX, Windows, and Macintosh platforms. Scilab has been developed since 1990 by researchers from the French National Institute for Research in Computer Science and Control (INRIA) and École Nationale des Ponts et Chaussées (ENPC). There is a multifractal–wavelet analysis toolbox available from the above website.

Mathematica (*http://www.wri.com/*). Mathematica seamlessly integrates a numeric and symbolic computational engine, graphics system, programming language, documentation system, and advanced connectivity to other applications. There is also a package called *Wavelet Explorer* from Wolfram Research. Wavelet Explorer generates a variety of orthogonal and biorthogonal filters and computes scaling functions, wavelets, and wavelet packets from a given filter. It also contains 1D and 2D wavelet and wavelet packet transforms, 1D and 2D local trigonometric transforms and packet transforms, and it performs multiresolution decomposition as well as 1D and 2D data compression and denoising tasks. Graphics utilities are provided to allow the user to visualize the results. It is written entirely in Mathematica with all source codes opened, enabling the user to customize and extend all the functions.

Maple (*http://www.maplesoft.com/*). Maple is symbolic computing software developed by researchers of Waterloo University since 1980. Scripts or programs in Maple can be easily generated to other languages, including MATLAB, FORTRAN, and C.

MathCAD (*http://www.mathsoft.com/*). MathCAD is another general purpose software platform with a symbolic computation engine for applying mathematics. It can work with mathematical expressions using standard math notation with the added ability to recalculate, view, present, and publish the data easily. MathCAD functionality can be extended through function extension packs and interoperable software packages, including add-ons dedicated to signal processing, image processing, wavelets, and solving and optimization. In particular, the Wavelets Extension Package provides one- and two-dimensional wavelets, discrete wavelet transforms, multiresolution analysis, and other tools for signal and image analysis, time-series analysis, statistical signal estimation, and data compression analysis.

MuPAD (*http://www.mupad.de/*). MuPAD is a computer algebra system originally developed by the MuPAD Research Group under the direction of Professor B. Fuchssteiner at the University of Paderborn, Germany. It is free (provided gratis) for personal use.

LastWave (*http://www.cmap.polytechnique.fr/~bacry/LastWave/index.html*). LastWave is a signal processing–oriented command language and is available for download from the Website. It is written in C and runs on X11/Unix, Windows/CygWin, and Macintosh platforms. It has been designed for use by anyone who knows about signal processing and wants to play around with wavelets and wavelet-like techniques. The command-line language employs a MATLAB-like syntax and includes high-level object-oriented graphic elements. It allows the user to deal with high-level structures such as signals, images, wavelet transforms, extrema representation, and short-time Fourier transform. One can add very easily some new commands in LastWave using either the command language itself or the C-language. New commands can be grouped into a package that can be loaded easily later. Several other packages have already been added to LastWave allowing high-level signal processing such as wavelet transforms (1D and 2D), extrema representations of wavelet transforms (1D and 2D), fractal analysis, matching pursuit, and compression.

Minitab (*http://www.minitab.com/*). Minitab is an all-in-one statistical and graphical analysis software package. Statistical tests such as regression analysis, analysis of variance, multivariate analysis, and time-series analysis can be completed easily. The Stat Guide is an extremely helpful tool in interpreting statistical graphs and analyses for users with minimal statistical background.

Extract (*http://www.extractinformation.com/*). Extract is designed for visualization of the structure and quantitative relationships in data. Multivariate statistical methods can then be used to model the data. A trial version of the software can be obtained after completion of a registration form on the Extract Information AB Webpage.

Sirius and Xtricator (*http://www.prs.no/*). Sirus and Xtricator are two pieces of software developed by the Pattern Recognition Systems (PRS), a company founded by Professor Olav M. Kvalheim of Norway. Sirius is a tool for multivariate analysis that covers all the essential chemometrics methods within multivariate exploratory analysis, classification/discrimination, and calibration/response modeling. Experimental design is also supported (please refer to Chapter 3 for an introduction to these methods). Xtricator is a software product that is developed for quantitative and qualitative analysis of data from hyphenated analytical instruments such as HPLC-DAD and GC-MS. It is capable of resolving two-way bilinear multicomponent instrumental data into spectra and chromatograms of pure analytes using heuristic evolving latent projection (HELP) techniques as mentioned in Section 3.6.2.3. A trial version of the software can be obtained after filling in a registration form on the PRS Webpage.

REFERENCES

1. T. Hirschfeld, *Anal. Chem.* **52**:297A–303A (1980).

2. R. Upton, ed., "Dang Gui Root, *Angelica sinensis* (Oliv.) Diels," in *Standards of Analysis, Quality Control, and Therapeutics*, American Herbal Pharmacopoeia, Scotts Valley, California, 2003.

3. F. Gong, F. T. Chau, and Y. Z. Liang, unpublished results.

4. S. Wold, "Chemometrics: What do we mean with it, and what do we want from it?" and S. D. Brown, "Has the 'chemometrics revolution' ended? Some views on the past, present and future of chemometrics," Papers on Chemometrics: Philosophy, History and Directions of InCINC '94, *http://www.emsl.pnl.gov:2080/docs/incinc/homepage.html*.

5. D. G. Howery and R. F. Hirsch, "Chemometrics in the chemistry curriculum," *J. Chem. Ed.* **60**:656–659 (1983).

6. CAMO ASA, Olav Tryggvasonsgt. 24, N-7011 Trondheim. Email: *camo@camo.no*.

7. PRS, Bergen High-Technology Center, Thormohlensgate 55, N-5008 Bergen, Norway. Email: *Info@prs.no*.

8. T. Braun, A. Schubert, and G. Schubert, "The most cited books in analytical chemistry," *Anal. Chem.* **73**:667A–669A (2001).

9. E. R. Malinowski, *Factor Analysis in Chemistry*, 3rd ed., Wiley, New York, 2002.

10. P. R. Bevington and D. K. Robinson, *Data Reduction and Error Analysis for the Physical Sciences*, 3rd ed., McGraw-Hill, New York, 2002.

11. N. R. Draper and H. Smith, *Applied Regression Analysis*, 2nd ed., Wiley, New York, 1981.

12. H. Martens and T. Naes, *Multivariate Calibration*, Wiley, Chichester, UK, 1989.

13. M. Otto, *Chemometrics: Statistics and Computer Applications in Analytical Chemistry*, Wiley, New York, 1999.

14. P. Geladi and B. R. Kowalski, "Partial least-squares regression: A tutorial," *Anal. Chim. Acta* **185**:1–17 (1986).

15. S. Wold, "Pattern recognition by means of disjoint principal components models," *Pattern Recogn.* **8**:127–139 (1976).

16. E. Vigneau, M. F. Devaux, E. M. Qannari, and P. Robert, "Principal component regression, ridge regression and ridge principal component regression in spectroscopy calibration," *J. Chemometr.* **11**:239–249 (1997).

17. X. H. Liang, J. E. Andrews, and J. A. Dehaseth, "Resolution of mixture components by target transformation factor analysis and determinant analysis for the selection of targets," *Anal. Chem.* **68**:378–385 (1996).

18. M. Maeder, "Evolving factor-analysis for the resolution of overlapping chromatographic peaks," *Anal. Chem.* **59**:527–530 (1987).

19. M. Maeder and A. Zilian, "Evolving factor-analysis, a new multivariate technique in chromatography," *Chemometr. Intell. Lab. Syst.* **3**:205–213 (1988).

20. C. N. Ho, G. D. Christian, and E. R. Davidson, "Application of the method of rank annihilation to quantitative analyses of multicomponent fluorescence data from the video fluorometer," *Anal. Chem.* **50**:1108–1113 (1978).

21. E. Sanchez and B. R. Kowalski, "Generalized rank annihilation factor analysis," *Anal. Chem.* **58**:499–501 (1986).

22. E. R. Malinowski, "Window factor-analysis—theoretical derivation and application to flow-injection analysis data," *J. Chemometr.* **6**:29–40 (1992).

23. E. R. Malinowski, "Automatic window factor analysis—a more efficient method for determining concentration profiles from evolutionary spectra," *J. Chemometr.* **10**:273–279 (1996).

24. O. M. Kvalheim and Y. Z. Liang, "Heuristic evolving latent projections—resolving 2-way multicomponent data. #1. Selectivity, latent-projective graph, datascope, local rank, and unique resolution," *Anal. Chem.* **64**:936–946 (1992).

25. Y. Z. Liang, O. M. Kvalheim, H. R. Kaller, D. L. Massart, P. Kiechle, and F. Erni, "Heuristic evolving latent projections—resolving 2-way multicomponent data. #2. Detection and resolution of minor constituents," *Anal. Chem.* **64**:946–953 (1992).

26. J. Zupan and J. Gasteiger, "Neural networks—a new method for solving chemical problems or just a passing phase," *Anal. Chim. Acta* **248**:1–30 (1991).

27. G. G. Andersson and P. Kaufmann, "Development of a generalized neural network," *Chemometr. Intell. Lab. Syst.* **50**:101–105 (2000).

28. D. B. Hibbert, "Genetic algorithms in chemistry," *Chemometr. Intell. Lab. Syst.* **19**:277–293 (1993).

29. C. B. Lucasius and G. Kateman, "Understanding and using genetic algorithms. #1. Concepts, properties and context," *Chemometr. Intell. Lab. Syst.* **19**:1–33 (1993).

30. C. B. Lucasius and G. Kateman, "Understanding and using genetic algorithms. #2. Representation, configuration and hybridization," *Chemometr. Intell. Lab. Syst.* **25**:99–145 (1994).

31. J. H. Kalivas, "Optimization using variations of simulated annealing," *Chemometr. Intell. Lab. Syst.* **15**:1–12 (1992).

32. U. Horchner and J. H. Kalivas, "Simulated-annealing-based optimization algorithms—fundamentals and wavelength selection applications," *J. Chemometr.* **9**:283–308 (1995).

33. S. G. Mallat, *A Wavelet Tour of Signal Processing*, Academic Press, London, 1998.

34. The MathWorks, Inc., *MATLAB for Windows*, Natick, Massachusetts, USA.

35. Y. Z. Liang, O. M. Kvalheim, and R. Manne, "White, grey, and black multicomponent systems. A classification of mixture problems and methods for their quantitative analysis," *Chemometr. Intell. Lab. Syst.* **18**:235–250 (1993).

36. K. M. Leung, F. T. Chau, and J. B. Gao, "Wavelet transform: A novel method for derivative calculation in analytical chemistry," *Anal. Chem.* **70**:5222–5229 (1998).

37. K. M. Leung, F. T. Chau, and J. B. Gao, "A review on applications of wavelet transform techniques in chemical analysis: 1989–1997," *Chemometr. Intell. Lab. Syst.* **43**:165–184 (1998).

38. X. G. Shao, A. K. M. Leung, and F. T. Chau, "Wavelet transform," *Acc. Chem. Research* **36**:276–283 (2003).

39. B. K. Alsberg, A. M. Woodward, and D. B. Kell, "An introduction to wavelet transform for chemometricians: A time-frequency approach," *Chemometr. Intell. Lab. Syst.* **37**:215–239 (1997).

CHAPTER

2

ONE-DIMENSIONAL SIGNAL PROCESSING
TECHNIQUES IN CHEMISTRY

In this chapter, some widely used methods for processing one-dimensional chemical signals are discussed. Chemical signals are usually recorded as chromatograms, spectra, voltammograms, kinetic curves, titration curves, and in other formats. Nowadays, almost all of them are acquired in the digital vector form. In order to denoise, compress, differentiate, and do other things on the signals acquired, it is always necessary to pretreat them first. The signal processing methods discussed in this chapter can be roughly divided into three classes including smoothing methods, transformation methods, and numerical treatment methods. They are described one by one below with illustrative examples.

2.1. DIGITAL SMOOTHING AND FILTERING METHODS

In general, averaging is widely used to improve the signal-to-noise ratio (SNR) for analytical signals. Through this treatment, the influence of noise can be reduced because the signals are often distributed normally on both positive and negative sides. In carrying out the averaging operation on a dataset x_i, the mean $\bar{x}$ can be calculated by

$$\bar{x} = \frac{1}{n} \sum_{i=1}^{n} x_i \tag{2.1}$$

Notice that the variance of $\bar{x}$ is only $1/\sqrt{n}$ of the original variable x_i. Thus, the averaging operation can increase the SNR of analytical signals. This explains why the spectrum of a sample generated from an infrared instrument in your laboratory is often the mean spectrum from several measurements. It should be pointed out here that most methods discussed in the following subsections are based on this principle.

Chemometrics: From Basics To Wavelet Transform, edited by Foo-tim Chau, Yi-zeng Liang, Junbin Gao, and Xue-guang Shao. Chemical Analysis Series, Vol. 164. ISBN 0-471-20242-8. Copyright © 2004 John Wiley & Sons, Inc.

2.1.1. Moving-Window Average Smoothing Method

The moving-window average method is the classic and the simplest smoothing method. It can be utilized to enhance the SNR. The principle of this method is illustrated in Figure 2.1. Suppose that we have a raw signal vector, say, $\mathbf{x} = [x_1, x_2, \ldots, x_{n-1}, x_n]$. In practice, a window size of $(2m+1)$ data points has to be specified first before doing any smoothing calculation. Here, an averaging filter of window size of $5(m = 2)$ is employed to illustrate the computing procedure. At the beginning, the first five data are used to find the first smoothed datum x_3^* via the following equation with $i = 3$ and $m = 2$:

$$x_i^* = \frac{1}{2m+1} \sum_{j=-m}^{m} x_{i+j} \qquad (2.2)$$

In this equation, x_i^* denotes the smoothed value while x_{i+j} are the original raw data, where i and j are the running indices. It should be noted that the first two data points, x_1 and x_2, cannot be smoothed in the process. After finding x_3^*, the next step is to move the window to the right by one datum to evaluate x_4^* (see Fig. 2.1). Then the procedure is repeated by moving the window successively along the equally spaced data until all the data are exhausted. As the width of the moving window is an important parameter

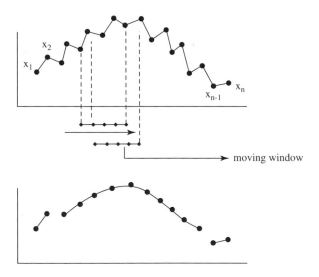

Figure 2.1. Moving-window-average filter for a window size of $2m+1 = 5$, that is, $m = 2$. It should be noted that for the extreme points, no smoothed data can be calculated because they are used for computing the first and the last averages. Top plot—the original raw signals; bottom plot—smoothed signals.

in this smoothing process, it has to be defined before any calculation is performed. The guidelines for choosing a right value for this parameter will be discussed in Section 2.3.

2.1.2. Savitsky–Golay Filter

The Savitsky–Golay filter is a smoothing filter based on polynomial regression [1–3]. Instead of simply using the averaging technique as mentioned previously, the Savitsky–Golay filter employs the regression fitting capacity to improve the smoothing results as depicted in Figure 2.2. From the plot, it can be seen that this method should perform better than the moving-window average method as mentioned in Section 2.1.1 because it takes advantage of the fitting ability of polynomial regression. However, the formulation of the Savitsky–Golay filter is quite similar to that of the averaging filter [Eq. (2.2)]. The major difference between the moving-window average method and the Savitsky–Golay filter is that the latter one is essentially a weighted average method in the form of

$$x_i^* = \frac{1}{2m+1} \sum_{j=-m}^{m} w_j x_{i+j} \tag{2.3}$$

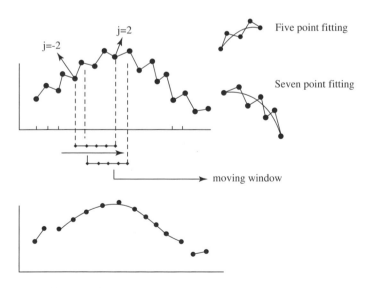

Figure 2.2. The Savitsky–Golay filter with a window size of $2m+1 = 5$. Instead of averaging the data point in the moving window, it uses the polynomial fitting technique to determine the weights in Equation (2.3). Top plot–the original raw signals; bottom plot–smoothed signals.

The problem now is how to find the correct weights w_j in this formula through polynomial regression. In the following example, a five-point Savitsky–Golay filter is utilized to show how this can be done.

The Savitsky–Golay filter uses the following formula to fit the data first (see Fig. 2.2), and then finds all the weights through the least-squares technique by

$$x_j^i = a_0 + a_1 j + a_2 j^2 + \cdots + a_k j^k$$

$$(j = -m, -m+1, \ldots, m-1, m; \ i = 1, \ldots, n) \qquad (2.4)$$

If the window size is 5 coupled with a quadratic model, formula (2.4) becomes

$$x_j^i = a_0 + a_1 j + a_2 j^2 \qquad (j = -2, -1, 0, 1, 2; \ i = 1, \ldots, n) \qquad (2.5)$$

where the superscript i indicates the numbering of the original chemical data and j is the window size variable. As x_j^i and j are known (see Fig. 2.2), the least-squares technique can be applied to find a_R ($R = 0, 1, 2$). By substituting the right values in Eq. (2.5), the following set of linear equations is obtained:

$$\begin{cases} x_{-2}^{i-2} = a_0 + a_1(-2) + a_2(-2)^2 \\ x_{-1}^{i-1} = a_0 + a_1(-1) + a_2(-1)^2 \\ x_0^i = a_0 + a_1(0) + a_2(0)^2 \\ x_1^{i+1} = a_0 + a_1(1) + a_2(1)^2 \\ x_2^{i+2} = a_0 + a_1(2) + a_2(2)^2 \end{cases} \qquad (2.6)$$

or

$$\begin{cases} x_{-2}^{i-2} = a_0 - 2a_1 + 4a_2 \\ x_{-1}^{i-1} = a_0 - a_1 + a_2 \\ x_0^i = a_0 \\ x_1^{i+1} = a_0 + a_1 + a_2 \\ x_2^{i+2} = a_0 + 2a_1 + 4a_2 \end{cases} = \begin{bmatrix} 1 & -2 & 4 \\ 1 & -1 & 1 \\ 1 & 0 & 0 \\ 1 & 1 & 1 \\ 1 & 2 & 4 \end{bmatrix} \begin{bmatrix} a_0 \\ a_1 \\ a_2 \end{bmatrix}$$

and in the matrix form

$$\begin{bmatrix} x_{-2}^{i-2} \\ x_{-1}^{i-1} \\ x_0^i \\ x_1^{i+1} \\ x_{+2}^{i+2} \end{bmatrix} = \begin{bmatrix} 1 & -2 & 4 \\ 1 & -1 & 1 \\ 1 & 0 & 0 \\ 1 & 1 & 1 \\ 1 & 2 & 4 \end{bmatrix} \begin{bmatrix} a_0 \\ a_1 \\ a_2 \end{bmatrix}$$

or

$$\mathbf{x} = \mathbf{Ma} \tag{2.7}$$

where

$$\mathbf{x} = \begin{bmatrix} x_{-2}^{i-2} \\ x_{-1}^{i-1} \\ x_0^i \\ x_1^{i+1} \\ x_{+2}^{i+2} \end{bmatrix}; \quad \mathbf{M} = \begin{bmatrix} 1 & -2 & 4 \\ 1 & -1 & 1 \\ 1 & 0 & 0 \\ 1 & 1 & 1 \\ 1 & 2 & 4 \end{bmatrix}; \quad \mathbf{a} = \begin{bmatrix} a_0 \\ a_1 \\ a_2 \end{bmatrix}$$

Equation (2.7) can be solved by least-squares fitting via

$$\mathbf{a} = (\mathbf{M}^t \mathbf{M}) \mathbf{M}^t \mathbf{x} \tag{2.8}$$

Substitution of the solution into Equation (2.7) gives the estimated values of $\hat{\mathbf{x}}$

$$\hat{\mathbf{x}} = \mathbf{M}(\mathbf{M}^t \mathbf{M}) \mathbf{M}^t \mathbf{x} \tag{2.9}$$

or in the common form

$$\begin{cases} \hat{x}_{-2}^{i-2} = \frac{1}{35}(31x_{-2}^{i-2} + 9x_{-1}^{i-1} - 3x_0^i - 5x_1^{i+1} + 3x_2^{i+2}) \\ \hat{x}_{-1}^{i-1} = \frac{1}{35}(9x_{-2}^{i-2} + 13x_{-1}^{i-1} + 12x_0^i + 6x_1^{i+1} - 5x_2^{i+2}) \\ \hat{x}_0^i = \frac{1}{35}(-3x_{-2}^{i-2} + 12x_{-1}^{i-1} + 17x_0^i + 12x_1^{i+1} - 3x_2^{i+2}) \\ \hat{x}_1^{i+1} = \frac{1}{35}(-5x_{-2}^{i-2} + 6x_{-1}^{i-1} + 12x_0^i + 13x_1^{i+1} + 9x_2^{i+2}) \\ \hat{x}_2^{i+2} = \frac{1}{35}(3x_{-2}^{i-2} - 5x_{-1}^{i-1} - 3x_0^i + 9x_1^{i+1} + 31x_2^{i+2}) \end{cases} \tag{2.10}$$

It should be mentioned that the Savitsky–Golay filter only utilizes the central point in the moving window to do smoothing. Thus, the weights $[w_{-2} \quad w_{-1} \quad w \quad w_1 \quad w_2]$ can be determined from the coefficients of x_0^i of the equations above, and they have values of $[-\frac{3}{35} \quad \frac{12}{35} \quad \frac{17}{35} \quad \frac{12}{35} \quad -\frac{3}{35}]$ for using five points in the moving window. One may also employ different window sizes and polynomials of different orders to deduce the weights. Just in this way, Savitsky and Golay collected the numbers with different window sizes in their tables (see Tables 2.1 and 2.2) for convenience in computation.

Example 2.1. An example using the Savitsky–Golay filter to smooth noisy analytical signals is given in Figure 2.3. The four plots in the figure show the original noisy signals and the signals that are smoothed by the Savitsky–Golay filter of quadratic/cubic polynomials with different window sizes. Four

Table 2.1. Weights of Savitsky–Golay Filter for Smoothing Based on a Quadratic/Cubic Polynomial

Points	25	23	21	19	17	15	13	11	9	7	5
−12	−253										
−11	−138	−42									
−10	−33	−21	−171								
−9	62	−2	−76	−136							
−8	147	15	9	−51	−21						
−7	222	30	84	24	−6	−78					
−6	287	43	149	89	7	−13	−11				
−5	343	54	204	144	18	42	0	−36			
−4	387	63	249	189	27	87	9	9	−21		
−3	422	70	284	224	34	122	16	44	14	−2	
−2	447	75	309	249	39	147	21	69	39	3	−3
−1	462	78	324	264	42	162	24	84	54	6	12
0	467	79	329	269	43	167	25	89	59	7	17
1	462	78	324	264	42	162	24	84	54	6	12
2	447	75	309	249	39	147	21	69	39	3	−3
3	422	70	284	224	34	122	16	44	14	−2	
4	387	63	249	189	27	87	9	9	−21		
5	343	54	204	144	18	42	0	−36			
6	287	43	149	89	7	−13	−11				
7	222	30	84	24	−6	−78					
8	147	15	9	−51	−21						
9	62	−2	−76	−136							
10	−33	−21	−171								
11	−138	−42									
12	−253										
	5,175	805	3,059	2,261	323	1,105	143	429	231	21	35

different window sizes were utilized to test the Savitsky–Golay filter. From the plots as given in Figure 2.3, one can see that as the window size increases, the smoothing effect becomes more significant. Yet, the tradeoff is that the distortion of the original signals become more serious. In this example, a window size of 13 seems to give the best result among all. Thus, the choice of window size for Savitsky–Golay filter is important.

To assist the reader in using the Savitsky–Golay filter for smoothing, a MATLAB source code is provided in the following frame:

```
function [y]=smoothing(x,win_num,poly_order)

%This is a program for smoothing the analytical signals.
%x can be a matrix with every column being an analytical
%signal vector.
```

```
%The parameter win_num is the window size which can be chosen
%to have a value of 7 to 17, say 7 9 11 13 15 17;
%The parameter poly_order is the polynomiar order which
%can be chosen to have a value of 2 or 3, and 4 or 5.

[m1,n1]=size(x);
y=zeros(size(x));
if win_num==7
if poly_order==2 | poly_order==3
  coef1=[-2 3 6 7 6 3 -2]/21;
for j=1:n1
for i=4:m1-3
  y(i,j)=coef1(1)*x(i-3,j)+coef1(2)*x(i-2,j)+coef1(3)*x(i-1,j)+ ...
         coef1(4)*x(i,j)+coef1(5)*x(i+1,j)+coef1(6)*x(i+2,j)+ ...
         coef1(7)*x(i+3,j);
end
end
else
coef1=[5 -30 75 131 75 -30 5]/231;
for j=1:n1
for i=4:m1-3
  y(i,j)=coef1(1)*x(i-3,j)+coef1(2)*x(i-2,j)+coef1(3)*x(i-1,j)+ ...
         coef1(4)*x(i,j)+coef1(5)*x(i+1,j)+coef1(6)*x(i+2,j)+ ...
         coef1(7)*x(i+3,j);
end
end
end
elseif win_num==9
  if poly_order==2|poly_order==3
coef1=[-21 14 39 54 59 54 39 14 -21]/231;
for j=1:n1
for i=5:m1-4
  y(i,j)=coef1(1)*x(i-4,j)+coef1(2)*x(i-3,j)+coef1(3)*x(i-2,j)+ ...
         coef1(4)*x(i-1,j)+coef1(5)*x(i,j)+coef1(6)*x(i+1,j)+ ...
         coef1(7)*x(i+2,j)+coef1(8)*x(i+3,j)+coef1(9)*x(i+4,j);
end
end
else
coef1=[15 -55 30 135 179 135 30 -55 15]/429;
for j=1:n1
for i=5:m1-4
  y(i,j)=coef1(1)*x(i-4,j)+coef1(2)*x(i-3,j)+coef1(3)*x(i-2,j) ...
         +coef1(4)*x(i-1,j)+coef1(5)*x(i,j)+coef1(6)*x(i+1,j) ...
         +coef1(7)*x(i+2,j)+coef1(8)*x(i+3,j)+coef1(9)*x(i+4,j);
end
end
end
elseif win_num==11

if poly_order==2|poly_order==3
coef1=[-36 9 44 69 84 89 84 69 44 9 -36]/429;
```

```
for j=1:n1
for i=6:m1-5
  y(i,j)=coef1(1)*x(i-5,j)+coef1(2)*x(i-4,j)+coef1(3)*x(i-3,j) ...
         +coef1(4)*x(i-2,j)+coef1(5)*x(i-1,j)+coef1(6)*x(i,j)+ ...
         coef1(7)*x(i+1,j)+coef1(8)*x(i+2,j)+coef1(9)*x(i+3,j) ...
         +coef1(10)*x(i+4,j)+coef1(11)*x(i+5,j);
end
end
else
coef1=[18 -45 -10 60 120 143 120 60 -10 -45 18]/429;
for j=1:n1
for i=6:m1-5
  y(i,j)=coef1(1)*x(i-5,j)+coef1(2)*x(i-4,j)+coef1(3)*x(i-3,j) ...
         +coef1(4)*x(i-2,j)+coef1(5)*x(i-1,j)+coef1(6)*x(i,j)+ ...
         coef1(7)*x(i+1,j)+coef1(8)*x(i+2,j)+coef1(9)*x(i+3,j) ...
         +coef1(10)*x(i+4,j)+coef1(11)*x(i+5,j);
end
end
end
elseif win_num==13
   if poly_order==2|poly_order==3
   coef1=[-11 0 9 16 21 24 25 24 21 16 9 0 -11]/143;
for j=1:n1
for i=7:m1-6
  y(i,j)=coef1(1)*x(i-6,j)+coef1(2)*x(i-5,j)+coef1(3)*x(i-4,j) ...
         +coef1(4)*x(i-3,j)+coef1(5)*x(i-2,j) ...
         +coef1(6)*x(i-1,j)+coef1(7)*x(i,j)+coef1(8)*x(i+1,j)+ ...
         coef1(9)*x(i+2,j)+coef1(10)*x(i+3,j)+coef1(11)*x(i+4,j) ...
         +coef1(12)*x(i+5,j)+coef1(13)*x(i+6,j);
end
end
else
coef1=[110 -198 -135 110 390 600 677 600 390 110 -135 -198 110]/2431;
for j=1:n1
for i=7:m1-6
  y(i,j)=coef1(1)*x(i-6,j)+coef1(2)*x(i-5,j)+coef1(3)*x(i-4,j) ...
         +coef1(4)*x(i-3,j)+coef1(5)*x(i-2,j)+coef1(6)*x(i-1,j) ...
         +coef1(7)*x(i,j)+coef1(8)*x(i+1,j)+ ...
         coef1(9)*x(i+2,j)+coef1(10)*x(i+3,j)+coef1(11)*x(i+4,j) ...
         +coef1(12)*x(i+5,j)+coef1(13)*x(i+6,j);
end
end
end
elseif win_num==15
   if poly_order==2|poly_order==3
   coef1=[-78 -13 42 87 122 147 162 167 162 147 122 87 42 -13 -78]/1105;
   for j=1:n1
for i=8:m1-7
  y(i,j)=coef1(1)*x(i-7,j)+coef1(2)*x(i-6,j)+coef1(3)*x(i-5,j) ...
```

```
               +coef1(4)*x(i-4,j)+coef1(5)*x(i-3,j)+coef1(6)*x(i-2,j) ...
               +coef1(7)*x(i-1,j)+coef1(8)*x(i,j)+coef1(9)*x(i+1,j) ...
               +coef1(10)*x(i+2,j)+coef1(11)*x(i+3,j)+coef1(12)*x(i+4,j) ...
               +coef1(13)*x(i+5,j)+coef1(14)*x(i+6,j)+coef1(15)*x(i+7,j);
end
end
else
  coef1=[2145 -2860 -2937 -165 3755 7500 10125 11063 10125 7500
         3755 -165 -2937 -2860 2145]/46189;
for j=1:n1
for i=8:m1-7
  y(i,j)=coef1(1)*x(i-7,j)+coef1(2)*x(i-6,j)+coef1(3)*x(i-5,j) ...
         +coef1(4)*x(i-4,j)+coef1(5)*x(i-3,j) ...
         +coef1(6)*x(i-2,j) ...
         +coef1(7)*x(i-1,j)+coef1(8)*x(i,j)+coef1(9)*x(i+1,j) ...
         +coef1(10)*x(i+2,j)+coef1(11)*x(i+3,j)+coef1(12)*x(i+4,j) ...
         +coef1(13)*x(i+5,j)+coef1(14)*x(i+6,j)+coef1(15)*x(i+7,j);
end
end
end
elseif win_num==17
if poly_order==2|poly_order==3
  coef1=[-21 -6 7 18 27 34 39 42 43 42 39 34 27 18 7 -6 -21]/323;
for j=1:n1
for i=9:m1-8
  y(i,j)=coef1(1)*x(i-8,j)+coef1(2)*x(i-7,j)+coef1(3)*x(i-6,j) ...
         +coef1(4)*x(i-5,j)+coef1(5)*x(i-4,j) ...
         +coef1(6)*x(i-3,j)+coef1(7)*x(i-2,j)+coef1(8)*x(i-1,j) ...
         +coef1(9)*x(i,j)+coef1(10)*x(i+1,j)+coef1(11)*x(i+2,j) ...
         +coef1(14)*x(i+5,j)+coef1(12)*x(i+3,j)+coef1(13)*x(i+4,j) ...
         +coef1(15)*x(i+6,j)+coef1(16)*x(i+7,j)+coef1(17)*x(i+8,j);
end
end
else
  coef1=[195 -195 -260 -117 135 415 660 825 883 825 660
         415 135 -117 -260 -195 195]/4199;
for j=1:n1
for i=9:m1-8
  y(i,j)=coef1(1)*x(i-8,j)+coef1(2)*x(i-7,j)+coef1(3)*x(i-6,j) ...
         +coef1(4)*x(i-5,j)+coef1(5)*x(i-4,j)+coef1(6)*x(i-3,j) ...
         +coef1(7)*x(i-2,j)+coef1(8)*x(i-1,j)+coef1(9)*x(i,j)+ ...
         coef1(10)*x(i+1,j)+coef1(11)*x(i+2,j)+coef1(12)*x(i+3,j) ...
         +coef1(13)*x(i+4,j)+coef1(14)*x(i+5,j) ...
         +coef1(15)*x(i+6,j)+coef1(16)*x(i+7,j)+coef1(17)*x(i+8,j);
end
end
end
end
```

Table 2.2. Weights of Savitsky–Golay Filter for Smoothing Based on a Quadratic/Cubic Polynomial

Points	25	23	21	19	17	15	13	11	9	7
−12	1,265									
−11	−345	95								
−10	−1,122	−38	11,628							
−9	−1,255	−95	−6,460	340						
−8	−915	−95	−13,005	−255	195					
−7	−255	−55	−11,220	−420	−195	2,145				
−6	590	10	−3,940	−290	−260	−2,860	110			
−5	1,503	87	6,378	18	−117	−2,937	−198	18		
−4	2,385	165	17,655	405	135	−165	−135	−45	15	
−3	3,155	235	28,190	790	415	3,755	110	−10	−55	5
−2	3,750	290	36,660	1,110	660	7,500	390	60	30	−30
−1	4,125	325	42,120	1,320	825	10,125	600	120	135	75
0	4,253	−339	44,003	1,393	883	11,063	677	143	179	131
1	4,125	325	42,120	1,320	825	10,125	600	120	135	75
2	3,750	290	36,660	1,110	660	7,500	390	60	30	−30
3	3,155	235	28,190	790	415	3,755	110	−10	−55	5
4	2,385	165	17,655	405	135	−165	−135	−45	15	
5	1,503	87	6,378	18	−117	−2,937	−198	18		
6	590	10	−3,940	−290	−260	−2,860	110			
7	−255	−55	−11,220	−420	−195	2,145				
8	−915	−95	−13,005	−255	195					
9	−1,255	−95	−6,460	340						
10	−1,122	−38	11,628							
11	−345	95								
12	1,265									
	30,015	6,555	260,015	7,429	4,199	46,189	2,431	429	429	231

2.1.3. Kalman Filtering

Kalman filtering is a kind of optimal linear recursive estimation method. Its operation speed is very high, and relatively small memory space is required for computation. Kalman filtering has been extensively used in engineering, especially in space technology. Recursive operation is the key feature of the method. Here we will first introduce what recursive operation is before discussing Kalman filtering in detail.

The basic idea of recursive operation is its efficient use of the results obtained previously and also the newly acquired information so as to avoid unnecessary repeated calculation. Let us first have a look at the basic feature of the recursive operation through a simple example. The mean

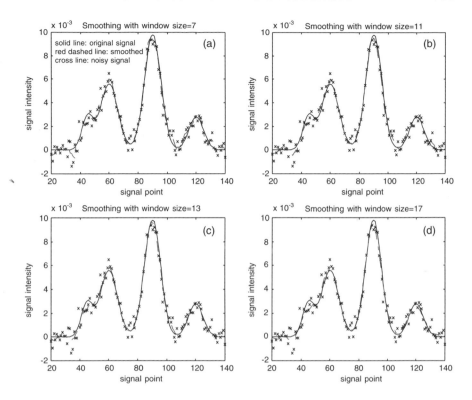

Figure 2.3. Smoothing results obtained by the Savitsky–Golay filter with different window sizes. They are depicted by four plots with the original curve (solid line), the raw noisy signals (cross line), and the smoothed curve (dashed line) with window size of 7 (a), 11 (b), 13 (c), and 17 (d).

value is usually evaluated using the following formula

$$\bar{x} = \frac{\Sigma x_i}{n} \tag{2.11}$$

where Σx_i denotes the sum of n observations, say x_i ($i = 1, \ldots, n$). When one measures a new x_i ($i = n + 1$), one has to calculate the mean again using Equation (2.11). Hence, all the n observations obtained before should be stored in the computer for future use. However, for recursive operation, a new mean can be evaluated through the following formula without using all the observations:

$$\bar{x}_{n+1} = \bar{x}_n + \frac{x_{n+1} - \bar{x}_n}{n+1} \tag{2.12}$$

Comparing this formula with Equation (2.11), one can obviously see that the recursive operation is faster and more efficient, and this is the attractive feature of Kalman filtering.

Kalman filter is based on a dynamic system model

$$\mathbf{x}(k) = \mathbf{F}(k, k-1)\mathbf{x}(k-1) + \mathbf{w}(k) \tag{2.13}$$

and a measurement model

$$y(k) = \mathbf{h}(k)^t \mathbf{x}(k-1) + e(k) \tag{2.14}$$

where $\mathbf{x}(k)$, $y(k)$, and $\mathbf{h}(k)$ denote the state vector, the measurement, and the measurement function vector, respectively. The variable k represents a measurement point that can be time, wavelength, or other. It should be noted that $\mathbf{F}(k, k-1)$ is the system transition matrix which represents how the system transits from state $(k-1)$ to state k. Very often, it is an identity matrix for smoothing purposes. $\mathbf{w}(k)$ denotes the dynamic system noise, and could be a zero vector approximately because the smoothing filter can be regarded as a static model. $e(k)$ is the measurement noise, which can be a stochastic variable with zero mean and constant variance obeying the Gaussian distribution.

The core recursive state estimate update in Kalman filtering is given by the following equation

$$\mathbf{x}(k) = \mathbf{x}(k-1) + \mathbf{g}(k)[\, y(k) - \mathbf{h}(k)^t \mathbf{x}(k-1)] \tag{2.15}$$

where the vector $\mathbf{g}(k)$ is called *Kalman gain*. Comparing this equation with Equation (2.12), one can easily see the similarity between the two. The Kalman gain, $\mathbf{g}(k)$, corresponds to $1/(n+1)$ in Equation (2.12) and is used to adjust the difference between the state vectors $\mathbf{x}(k)$ and $\mathbf{x}(k-1)$ through the term of measurement difference, of $[y(k) - \mathbf{h}(k)^t \mathbf{x}(k-1)]$. Through Equation (2.15), one can also see that the state estimate update is just based on the newly measured $y(k)$ and the state vector $\mathbf{x}(k-1)$ obtained before. Equation (2.15) makes the efficient usage of recursive operation possible.

The Kalman gain can be determined by the following formula

$$\mathbf{g}(k) = \mathbf{P}(k-1)\mathbf{h}(k)[\mathbf{h}(k)^t \mathbf{P}(k-1)\mathbf{h}(k) + r(k)]^{-1} \tag{2.16}$$

where $r(k)$ represents the variance of the measurement noise $e(k)$. $\mathbf{P}(k-1)$ is the covariance matrix of the system estimated from the $(k-1)$ observations obtained before through

$$\mathbf{P}(k) = [\mathbf{I} - \mathbf{g}(k-1)\mathbf{h}(k)^t]\mathbf{P}(k-1)[\mathbf{I} - \mathbf{g}(k-1)\mathbf{h}(k)^t]$$
$$+ \mathbf{g}(k-1)r(k)\mathbf{g}(k-1)^t \tag{2.17}$$

where $\mathbf{I}$ is an identity matrix.

From the discussion above, it can be seen that the Kalman gain vector can be deduced through Equation (2.16) if the initial values of $\mathbf{x}(k)$ and $\mathbf{P}(k)$, say, $\mathbf{x}(0)$ and $\mathbf{P}(0)$, are known. Then, the next $\mathbf{x}(k)$ and $\mathbf{P}(k)$ can be computed through Equations (2.15) and (2.17) until convergence is attained.

In summary, the procedure of Kalman filtering can be carried out via the following steps:

1. Setting the initial values:

$$\mathbf{x}(0) = \mathbf{0}, \quad \mathbf{P}(0) = \gamma^2 \mathbf{I} \tag{2.18}$$

where γ^2 is an initial estimation of variance of measurement noises that might be given by the following empirical formula

$$\gamma^2 = a\frac{r(1)}{[\mathbf{h}(1)^t\mathbf{h}(1)]^{1/2}} \tag{2.19}$$

The factor a can influence the calculation accuracy and can have values from 10 to 100. It is worthwhile to note that the initial value of $\mathbf{P}(0)$ is crucial for the estimation. If its value is too small, it can result in bias estimation. Yet, if its value is too high, it is difficult to have the computation converging to the desired value.

2. Recursive calculation loop:

$$\mathbf{g}(k) = \mathbf{P}(k-1)\mathbf{h}(k)[\mathbf{h}(k)^t\mathbf{P}(k-1)\mathbf{h}(k) + r(k)]^{-1}$$
$$\mathbf{x}(k) = \mathbf{x}(k-1) + \mathbf{g}(k)[\,y(k) - \mathbf{h}(k)^t\mathbf{x}(k-1)]$$
$$\mathbf{P}(k) = [\mathbf{I} - \mathbf{g}(k-1)\mathbf{h}(k)^t]\mathbf{P}(k-1)[\mathbf{I} - \mathbf{g}(k-1)\mathbf{h}(k)^t]$$
$$+ \mathbf{g}(k-1)r(k)\mathbf{g}(k-1)^t$$

where $r(k)$ is the variance of measurement noises that can be determined by the variance of real noise. This loop procedure is repeated until the estimates become stable.

In Kalman filtering algorithm, the innovative series is very important and might provide information about whether the results obtained are reliable. The innovative series can be obtained by the following equation:

$$v(k) = y(k) - \mathbf{h}(k)^t\mathbf{x}(k-1) \tag{2.20}$$

In fact, the series is the difference between the measurement and estimation and can be regarded as a residual at the k point. The innovative series should be a white noise with zero mean if the filtering model used is correct. Otherwise, the results obtained are not reliable.

Kalman filtering can be applied for filtering, smoothing, and prediction. The most common application is known in multicomponent analysis.

2.1.4. Spline Smoothing

In addition to the smoothing methods based on digital filters as discussed previously, the other widely used one in signal processing is spline functions. The main advantage of spline functions is their differentiability in the entire measurement domain.

Among various spline functions, the cubic spline function is the most common one and is defined as follows

$$y = S(x) = A_k(x - x_k)^3 + B_k(x - x_k)^2 + C_k(x - x_k) + D_k \qquad (2.21)$$

where A_k, B_k, C_k, and D_k are the spline coefficients at data point k. The cubic spline function $S(x)$ or y for observations on the abscissa intervals $x_1 < x_2 < \cdots < x_n$ satisfies the following conditions:

1. The intervals are called *knots*. The knots may be identical with the index points on the x axis (abscissa).
2. Within the knots k, $S(x)$ obeys the continuity constraint on the function and on its twofold derivatives.
3. $S(x)$ is a cubic function in each subrange $[x_k, x_{k-1}]$ for $k = 1, \ldots, n-1$ considered.
4. Outside the range from x_1 to x_k, $S(x)$ is a straight line.

For a fixed interval between the data points x_k and x_{k-1}, the following relationships are valid for the signal values and their derivatives:

$$y_k = D_k$$
$$y_{k+1} = A_k(x - x_k)^3 + B_k(x - x_k)^2 + C_k(x - x_k)$$
$$y_k' = S'(x_k) = C_k$$
$$y_{k+1}' = 3A_k(x - x_k)^2 + 2B_k(x - x_k) + C_k$$
$$y_k'' = S''(x_k) = 2B_k$$
$$y_{k+1}'' = 6A_k(x - x_k) + 2B_k$$

The spline coefficients can be determined by a method that also smoothes the data under study at the same time. The ordinate values $\hat{y}_k$ are calculated such that the differences of the observed values are positive proportional jumps r_k in their third derivative at point x_k:

$$r_k = S'''(x_k) - S'''(x_{k+1}) \qquad (2.22)$$

$$r_k = p_k(y_k - \hat{y}_k) \qquad (2.23)$$

The proportionality factors p_k are determined by cross-validation. In contrast with polynomials, spline functions may be applied to approximate and smooth any kind of curve shape. It should be mentioned that many more coefficients must be estimated and stored in comparison with the polynomial filters because different coefficients apply in each interval. A disadvantage is valid for smoothing splines where the parameter estimates are biased. Therefore, it is more difficult to describe the statistical properties of spline functions than those of linear regression.

In MATLAB, there is a cubic spline function, named *csaps*. csaps(X, Y, p, X), which returns a smoothed version of the input data (X, Y) by cubic smoothing spline, and the result depends on the value of the smoothing parameter p (from 0 to 1). For $p = 0$, the smoothing spline corresponds to the least-squares straight-line fit to the data, while at the other extreme, with $p = 1$, it is the "natural" or variational cubic spline interpolation. The transition region between these two extremes is usually only a rather small range of values for p and its optimal value strongly depends on the nature of the data. Figure 2.4 shows an example of smoothing by a cubic spline smoother with different p values. From the plots as given in the figure, one can see that the choice of the right value for parameter p is crucial. The smoothing results are satisfactory if one makes a good choice as depicted in Figure 2.4c. In order to make it easier for the readers to understand the smoothing procedure using the cubic spline smoother, a MATLAB source code is given in the following frame:

```
xi=[0:.05:1.5];
yi=cos(xi)
ybad=yi+.2*(rand(size(xi))-.5);
figure(2)
subplot(221),plot(xi,yi,'k:',xi,ybad,'kx'),grid on
title('Original curve: dashed line; Noisey data: cross')
axis([0 1.5 0 1.2])
xlabel('Varibale (x)')
ylabel('Signal, (y)')

yy1=csaps(xi,ybad,.9981,xi);
subplot(222),plot(xi,yi,'k:',xi,ybad,'kx',xi,yy1,'k'), grid on
title('Smoothed curve: solid line with p=9981')
axis([0 1.5 0 1.2])
```

```
xlabel('Varibale (x)')
ylabel('Signal, (y)')

yy2=csaps(xi,ybad,.9756,xi);
subplot(223),plot(xi,yi,'k:',xi,ybad,'kx',xi,yy2,'k'), grid on
title('Smoothed curve: solid line with p=9756')
axis([0 1.5 0 1.2])
xlabel('Varibale (x)')
ylabel('Signal, (y)')
yy3=csaps(xi,ybad,.7856,xi);
subplot(224),plot(xi,yi,'k:',xi,ybad,'kx',xi,yy3,'k'), grid on
title('Smoothed curve: solid line with p=7856')
axis([0 1.5 0 1.2])
xlabel('Varibale (x)')
ylabel('Signal, (y)')
```

Usually, it is difficult to choose the best value for the parameter p without experimentation. If one has difficulty in doing this but has an idea of the noise level in Y, the MATLAB command spaps(X, Y, tol) may help. Select

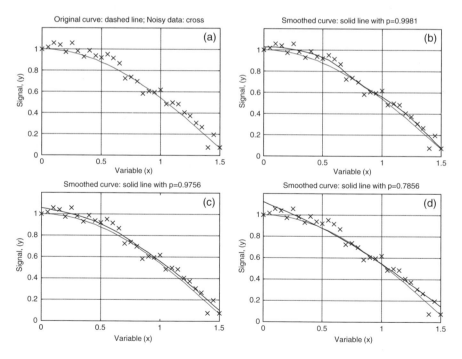

Figure 2.4. Smoothing results obtained by a cubic spline smoother with different values of the parameter p: (a) the original curve and the raw noisy signals; (b) the smoothed curve with $p = 0.9981$; (c) the smoothed curve with $p = 0.9756$; (d) the smoothed curve with $p = 0.7856$.

a p value in such a way that

$$\text{tol} = \sum_i (y_i - \hat{y}_i)^2 \qquad (2.24)$$

so as to produce the smoothest spline within an acceptable tolerance for the data.

2.2. TRANSFORMATION METHODS OF ANALYTICAL SIGNALS

Transformation is a very useful technique in pretreatment of analytical signals. Convolution, Hadamard, and Fourier transformation are just examples of this kind. In essence, wavelet analysis is also another kind of transformation technique. In this section the methods of convolution, Hadamard and Fourier transformation will be discussed in some detail.

2.2.1. Physical Meaning of the Convolution Algorithm

Convolution plays a very important role in statistics in treating analytical signals. An example from spectral measurement is now presented to illustrate how convolution works [6].

Suppose that the real spectrum of a sample is the one given by $f(x)$ in the upper part of Figure 2.5. Now, a spectrometer with a slit is utilized to assist the acquisition of the spectrum. If the slit is infinitely narrow, the spectral signal recorded should be the same as that of $f(x)$. In practice, any slit has certain width. Let the slit operation be expressed by function $h(x)$ which is essential a triangular function (see the lower part of Fig. 2.5). From the plot of $h(x)$, one can see how the slit function (triangular function) affects the intensity distribution of the lightbeam with respect to the location. The highest-intensity location appears at the central point of the slit. Thus, when the light ray passes this slit, the spectrum obtained [as expressed by the function $g(x)$ in Fig. 2.5] by the spectrophotometer becomes broader than the real one $f(x)$. The whole procedure as described above is called *convolution* in the field of signal processing.

From this plot (Fig. 2.5), it can be seen that the slit works somewhat like the Savitsky–Golay filter. The triangular function is essentially a weight function. That is why the Savitsky–Golay filter is originally called the *polynomial convolution method*. Since the spectrum $g(x)$ obtained from the spectrophotometer is the convolution result of the original spectrum $f(x)$ and the triangular function $h(x)$, the term $g(x)$ can be expressed

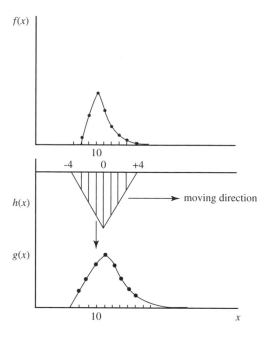

Figure 2.5. Illustration of the physical meaning of convolution by the slit function.

mathematically by the following formula:

$$g[x(i)] = \sum_{i=-m}^{m} f(x) \cdot h[x(i) - x] \tag{2.25}$$

This formula is the discrete expression of the convolution operation through which one can see that $N = 2m + 1$ is the width of the slit. In Equation (2.25), $x(i)$ represents the intensity of the light of the measured spectrum at the central point. It should be noted that all the elements of the slit function $h(x)$ outside the moving window have zero values. Thus, the continuous formula of convolution can be expressed as follows:

$$g[x(i)] = \int_{-\infty}^{+\infty} f(x)h(x(i) - x)dx \tag{2.26}$$

Let $x(i)$ be represented by y. Then we have

$$g(y) = \int_{-\infty}^{+\infty} f(x)h(y - x)dx \tag{2.27}$$

Here $g(y)$ is usually called the *convolution* of functions $f(y)$ and $h(y)$ and is denoted by $f(y) * h(y)$.

It should be mentioned that the convolution operation can be easily fulfilled by the Fourier transformation. Hence, the convolution operation is essentially a kind of transformation.

2.2.2. Multichannel Advantage in Spectroscopy and Hadamard Transformation

The major advantage of multichannel measurement in spectroscopic study is illustrated by the following design in an weighting experiment [7]. Suppose that there are four alloy samples to be weighed. The variance of weighting is σ^2 if they are measured one by one in the usual practice. However, the weighting experiment can be designed in another way by putting different combinations of the four samples on the two sides of a balance. From these measurements, the following relationship can be established:

$$\begin{cases} m_1 = x_1 + x_2 + x_3 + x_4 + e_1 \\ m_2 = x_1 - x_2 - x_3 + x_4 + e_2 \\ m_3 = x_1 - x_2 + x_3 - x_4 + e_3 \\ m_4 = x_1 + x_2 - x_3 - x_4 + e_4 \end{cases} \tag{2.28}$$

Here m_i ($i = 1, 2, 3, 4$) denotes the weight on the left-hand side of the balance. As for x_i ($i = 1, 2, 3, 4$), it represents the weight of the alloy sample of which the one with the minus sign means that it is placed on the left-hand side of the balance and the one with the positive sign is on the right hand side. From the preceding linear equations, one can easily obtain the estimation of x_i. For instance, the estimation of x_1 is given by

$$\hat{x}_1 = \frac{1}{4}(m_1 + m_2 + m_3 + m_4) = x_1 + \frac{1}{4}\sum e_i \tag{2.29}$$

From this result, it can be seen that the error is only one-fourth that from the usual practice, or in another words, the variance for this weighting design is $\sigma^2/16$. This example illustrates that a smart design can improve the accuracy of weighing significantly.

Let us elaborate the weighting experiment above in more detail from the mathematical point of view. If the four samples are weighted one by one, the result obtained can be expressed by the following equations (ignoring the error term)

$$\begin{bmatrix} m_1 \\ m_2 \\ m_3 \\ m_4 \end{bmatrix} = \begin{bmatrix} 1 & 0 & 0 & 0 \\ 0 & 1 & 0 & 0 \\ 0 & 0 & 1 & 0 \\ 0 & 0 & 0 & 1 \end{bmatrix} \begin{bmatrix} x_1 \\ x_2 \\ x_3 \\ x_4 \end{bmatrix} \tag{2.30}$$

or in matrix form as $\mathbf{m} = \mathbf{Ax}$,

$$\mathbf{x} = \mathbf{A}^{-1}\mathbf{m} = \mathbf{Am}$$

where $\mathbf{A}$ is essentially an identity matrix. The calculation procedure is very simple, but every sample is weighted only once. With the use of the smart weighting procedure, the outcomes [Eq. (2.28)] can be expressed as

$$\begin{bmatrix} m_1 \\ m_2 \\ m_3 \\ m_4 \end{bmatrix} = \begin{bmatrix} 1 & 1 & 1 & 1 \\ 1 & -1 & -1 & 1 \\ 1 & -1 & 1 & -1 \\ 1 & 1 & -1 & -1 \end{bmatrix} \cdot \begin{bmatrix} x_1 \\ x_2 \\ x_3 \\ x_4 \end{bmatrix} \tag{2.31}$$

or $\mathbf{m} = \mathbf{Hx}$. Matrix $\mathbf{H}$ is known as the *Hadamard matrix*. Obviously, the absolute value of every element, H_{mn}, in $\mathbf{H}$ is 1:

$$|H_{mn}| = 1 \tag{2.32}$$

In this case, the inverse matrix of $\mathbf{H}$ is also very simple:

$$\mathbf{H}^{-1} = \frac{1}{N}\mathbf{H} \tag{2.33}$$

The approach of the smart weighting design can be applied to spectral analysis to improve its performance. In traditional spectral measurement, the intensity of each wavelength is measured one at a time to obtain the spectrum. Figure 2.6 depicts the operation of such a single-slit scanning spectrometer. But if the intensities of N different wavelengths are recorded simultaneously, the signal-to-noise ratio can be enhanced significantly. According to the weighting design experiment above, one needs to take only two values, say, $+1$ and 0. This can be realized easily in spectral measurement. Here the zero value means that there is no light passing through, while 1 means "Yes." Suppose that one uses a rather wide slit

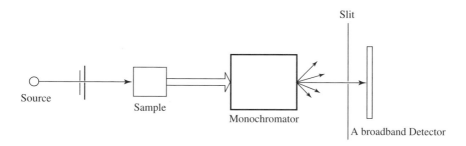

Figure 2.6. Schematic diagram of a single-slit scanning spectrometer.

compared to the wavelength under study and tries to cover some wavelengths. For instance, the light intensities of seven different wavelengths ψ_j ($j = 1, 2, \ldots, 7$) are acquired at one time. The traditional measurement method will take them one by one. Yet, we can also design a spectrophotometer following the preceding smart weighting design by taking several selected wavelengths simultaneously. Let $1 \times \psi_j$ denote the light ray that can pass through the slit, and assume that $0 \times \psi_j$ means that it cannot. Then, for a measurement using the design in the series of 1001011, the total light intensity detected is expressed as follows:

$$\psi = 1 \times \psi_1 + 0 \times \psi_2 + 1 \times \psi_3 + 1 \times \psi_4 + 0 \times \psi_5 + 1 \times \psi_6 + 1 \times \psi_7$$
$$= \psi_1 + \psi_4 + \psi_6 + \psi_7$$

Anyone who is clever enough to design seven independent combinations of spectral measurements can easily obtain the seven individual ψ_j correctly. One design of this kind is given in the following matrix:

$$\mathbf{S} = \begin{bmatrix} 1 & 0 & 0 & 1 & 0 & 1 & 1 \\ 1 & 1 & 0 & 0 & 1 & 0 & 1 \\ 1 & 1 & 1 & 0 & 0 & 1 & 0 \\ 0 & 1 & 1 & 1 & 0 & 0 & 1 \\ 1 & 0 & 1 & 1 & 1 & 0 & 0 \\ 0 & 1 & 0 & 1 & 1 & 1 & 0 \\ 0 & 0 & 1 & 0 & 1 & 1 & 1 \end{bmatrix} \tag{2.34}$$

This is called the *Sylvester matrix*, and it can be derived from the Hadamard matrix. The Sylvester matrix is obtained in the following way. First, set the elements of the first row. Then the second row is finished by moving the last element in the first row of the matrix to become the first element. Afterward, move the remaining six elements sequentially to the right by one position. The elements of the third and other rows are expressed in the same way, based on the previous rows. This kind of measurement, called the *Hadamard coding procedure* in spectral study, is utilized for Hadamard transformation spectroscopy. A schematic diagram of the Hadamard multichannel spectrometer is shown in Figure 2.7.

It should be mentioned that Hadamard transformation spectroscopy can enhance the signal-to-noise ratio (SNR) by $(N + 1)/2N^{1/2}$ times. When the value of N is large enough, the increase in SNR can reach $N^{1/2}/2$ times [8].

This multichannel advantage in spectroscopic analysis by using Hadamard transformation as mentioned above is called the *Fellgett advantage*. In fact, Hadamard transformation infrared spectroscopy has been found to enhance SNR notably.

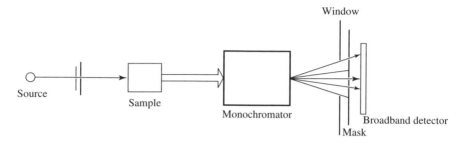

Figure 2.7. Schematic diagram of the Hadamard multichannel spectrometer.

2.2.3. Fourier Transformation

Fourier transformation is a widely used mathematical technique [9–12] for converting a signal, $f(t)$, from the time domain into the frequency domain, $F(\nu)$. This is because almost all the continuous signals can be expressed by the periodic trigonometric functions, sine and cosine functions, in the form of

$$f(t) = a_0 + \sum_{n=1}^{\infty} a_n \cos(nw_0 t) + b_n \sin(nw_0 t) \qquad (2.35)$$

where $f(t)$ is a function in the time domain and $w_0 = 2\pi f_0$ with $f_0 = 1/t$ and a_0 represents the so-called direct-current term, which can be deduced by a mean value in a period (T) of $x(t)$:

$$a_0 = \frac{1}{T} \int_0^T f(t)dt \qquad (2.36)$$

The other coefficients, a_n and b_n ($n = 1, 2, 3, \ldots$), in this expression are represented by the following expressions:

$$a_n = \frac{2}{T} \int_0^T f(t) \cos(nw_0 t)dt \qquad (2.37a)$$

$$b_n = \frac{2}{T} \int_0^T f(t) \sin(nw_0 t)dt \qquad (n = 1, 2, 3, \ldots) \qquad (2.37b)$$

Fourier transformation is essentially a kind of frequency analysis, which is quite similar to the procedure of splitting a lightbeam from a source into different light rays at different wavelengths by a prism or grating.

Let $\cos(2\pi n f_0 t)$ and $\sin(2\pi n f_0 t)$ be expressed in the following way:

$$\cos(2\pi n f_0 t) = \frac{e^{j2\pi n f_0 t} + e^{-j2\pi n f_0 t}}{2} \tag{2.38a}$$

$$\sin(2\pi n f_0 t) = \frac{e^{j2\pi n f_0 t} - e^{-j2\pi n f_0 t}}{2j} \tag{2.38b}$$

Then the Fourier series can be written as

$$f(t) = \sum_{n=-\infty}^{\infty} C_n e^{j2\pi n f_0 t} \tag{2.39}$$

Here C_n are the values taken with n as $-\infty, \ldots, -1, 0, 1, 2, \ldots, \infty$, respectively

$$C_n = \frac{1}{T} \int_{-\frac{T}{2}}^{\frac{T}{2}} f(t) e^{-j2\pi n f_0 t} \, dt \tag{2.40}$$

For every n, C_n will essentially give the swing and phase of the homorous wave with a frequency $f = n f_0$.

The formal definition of Fourier transformation is given below. Let a function $f(t)$ be in the time or space domain. Its expression of Fourier transformation will be

$$f(f) = \int_{-\infty}^{\infty} f(t) e^{-j2\pi f_0 t} \, dt \tag{2.41}$$

The inverse Fourier transformation is defined as follows:

$$f(t) = \int_{-\infty}^{\infty} f(f) e^{j2\pi f} \, df \tag{2.42}$$

Thus, $f(f)$ can be converted back to $f(t)$ through the preceding formula. This means that the function can be freely exchanged between the time or space domains and the frequency domain through Eqs. (2.41) and (2.42):

$$f(t) \Leftrightarrow f(f)$$

2.2.3.1. Discrete Fourier Transformation and Spectral Multiplex Advantage

In general, we often use the discrete Fourier transformation to pretreat chemical measurements. Suppose that the time-domain signal $f(t)$ is sampled at N equally intervals. This is called the *discrete Fourier transform*

(DFT):

$$f(n) = \sum_{k=0}^{n-1} f(k)e^{-j2\pi k(n/N)} \tag{2.43}$$

For inverse transformation, we have

$$f(k) = \frac{1}{N} \sum_{n=0}^{N-1} f(n)^{j2\pi k(n/N)} \tag{2.44}$$

It can be easily seen that if one can sample the signals by N equally spaced intervals using a multichannel detector (see Fig. 2.8), then N data points $f(t_k)$ or $f(k)$ in the time or space domain, where $k = 1, 2, \ldots, N$, can be obtained. For every such data point acquired, one can get the corresponding series of frequency domain amplitudes, say, $f(f_n)$ or $f(n)$, with the help of DFT. Therefore, the spectral multiplex advantage of Hadamard transformation as discussed in the previous section also happens in Fourier transformation. For the Hadamard transformation, we have $\mathbf{m} = \mathbf{Hx}$. In the same way, the Fourier transformation matrix $\mathbf{F}$ can be employed to accomplish the spectral multiplex advantage: $\mathbf{m} = \mathbf{Fx}$. The elements F_{mn} in matrix $\mathbf{F}$ can be expressed as $F_{mn} = \exp(2\pi jmn/N) = \cos(2\pi mn/N) + j\sin(2\pi mn/N)$ with $j = \sqrt{-1}$. It is easily seen that $|F_{mn}| = 1$. Letting $N = 4$, we have

$$\begin{bmatrix} m_1 \\ m_2 \\ m_3 \\ m_4 \end{bmatrix} = \begin{bmatrix} 1 & 1 & 1 & 1 \\ 1 & j & -1 & j \\ 1 & -1 & 1 & -1 \\ 1 & -j & -1 & j \end{bmatrix} \begin{bmatrix} x_1 \\ x_2 \\ x_3 \\ x_4 \end{bmatrix} \tag{2.45}$$

It should be noted that only some channels of the spectral signals are utilized in Hadamard transformation while all the spectral signals are considered in DFT (see Figs. 2.7 and 2.8 for comparison). In this way, SNR can be enhanced by $N^{1/2}$.

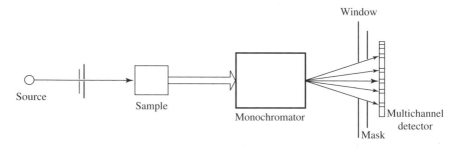

Figure 2.8. Schematic diagram of a multichannel-detector-based spectrometer.

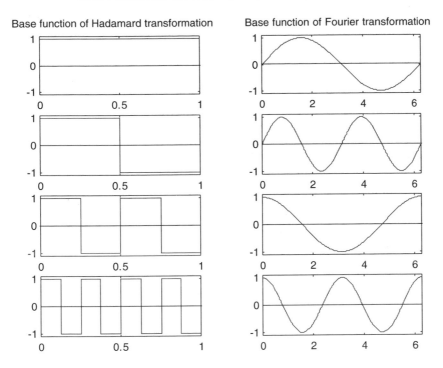

Figure 2.9. Base function of Hadamard (left part) and Fourier transformation (right part).

Hadamard transformation and Fourier transformation differ from each other in the base function. Hadamard transformation is based on the Walsh function, in contrast to the sine and cosine functions in Fourier transformation. This is illustrated in Figure 2.9.

Example 2.2. A signal in the time domain can be represented by a combination of periodic sine and cosine functions. Usually, any time-dependent or continuous signal can be considered as a combination of sine and cosine functions. This explains why Fourier transformation has wide application.

Figure 2.10 illustrates how Fourier transformation works. The plot shown in Figure 2.10a is the sum of three trigonometric functions (Fig. 2.10c) with two sine functions with the periods of 1π and 2π as well as one cosine function with the period of 3π. Through applying Fourier transformation to the plot, the dependence of the intensity on frequency from the calculation is depicted in Figure 2.10b. It can be seen from the figure that the three component functions are definitely identified.

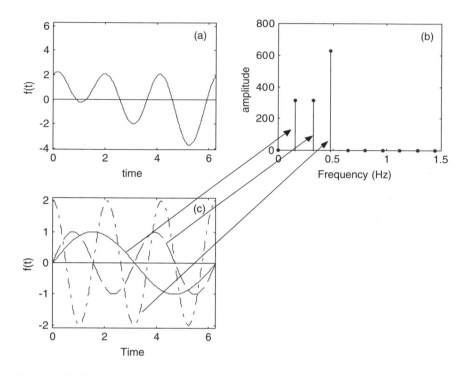

Figure 2.10. Fourier transformation: the composite signal (a) contains two sine functions with the periods of 1π and 2π and one cosine function with the period of 3π (b); the dependence of intensity on frequency after Fourier transformation (c).

2.2.3.2.   *Fast Fourier Transformation*

In this section we briefly describe how the fast Fourier transformation can be used to carry out inverse Fourier transformation. For more detail, readers can refer to Brigham's treatise [11]. Here a simple case of $N = 4$ is considered. Let us define w to be a complex number

$$w = e^{-2\pi j/4} \tag{2.46}$$

Then, the expression of DFT can be written as

$$f(n) = \sum_{k=0}^{N-1} f(k)w^{nk} \tag{2.47}$$

The basic idea of fast Fourier transform (FFT) is to decompose this formula so as to reduce the calculation burden. When N is equal to a power of 2, that is, $N = 2^a$ where a is an integer, the computation is very simple. Now, $N = 2^2 = 4$ is utilized as an example to illustrate the FFT decomposition procedure [18, 19]. In this case, Equation (2.47) becomes

$$f(n) = \sum_{k=0}^{4-1} f_0(k)w^{nk} \qquad n = 0, 1, 2, 3 \qquad (2.48)$$

A slight change in the notation of $f(k)$ to $f_0(k)$ is made here so as to indicate the signal before Fourier transformation. This equation can be written as follows:

$$\begin{cases} f(0) = f_0(0)w^0 + f_0(1)w^0 + f_0(2)w^0 + f_0(3)w^0 \\ f(1) = f_0(0)w^0 + f_0(1)w^1 + f_0(2)w^2 + f_0(3)w^3 \\ f(2) = f_0(0)w^0 + f_0(2)w^2 + f_0(2)w^4 + f_0(3)w^6 \\ f(3) = f_0(0)w^0 + f_0(1)w^3 + f_0(2)w^6 + f_0(3)w^9 \end{cases} \qquad (2.49)$$

or

$$\begin{bmatrix} f(0) \\ f(1) \\ f(2) \\ f(3) \end{bmatrix} = \begin{bmatrix} w^0 & w^0 & w^0 & w^0 \\ w^0 & w^1 & w^2 & w^3 \\ w^0 & w^2 & w^4 & w^6 \\ w^0 & w^3 & w^6 & w^9 \end{bmatrix} \begin{bmatrix} f_0(0) \\ f_0(1) \\ f_0(2) \\ f_0(3) \end{bmatrix} \qquad (2.50)$$

In matrix form, we have

$$f(n) = \mathbf{W}^{nk} f_0(k)$$

where

$$f(n) = \begin{bmatrix} f(0) \\ f(1) \\ f(2) \\ f(3) \end{bmatrix}; \qquad \mathbf{W}^{nk} = \begin{bmatrix} w^0 & w^0 & w^0 & w^0 \\ w^0 & w^1 & w^2 & w^3 \\ w^0 & w^2 & w^4 & w^6 \\ w^0 & w^3 & w^6 & w^9 \end{bmatrix} \qquad \text{and} \qquad f_0(k) = \begin{bmatrix} f_0(0) \\ f_0(1) \\ f_0(2) \\ f_0(3) \end{bmatrix}$$

It should be noted that $w^{nk} = e^{-2\pi j(nk/4)}$. Here the symbol ξ is used to denote the remainder of the division of (kn) by 4. Then, only ξ is needed to be considered in the following calculation. For $nk = 6$, we have

$$w^6 = e^{-j2\pi 6/4} = e^{-j2\pi(4/4)}e^{-j2\pi 2/4} = e^{-(j2\pi/4)2} = w^2$$

and $\xi = 2$. Equation (2.50) can now be written as

$$\begin{bmatrix} f(0) \\ f(1) \\ f(2) \\ f(3) \end{bmatrix} = \begin{bmatrix} 1 & 1 & 1 & 1 \\ 1 & w^1 & w^2 & w^3 \\ 1 & w^2 & w^0 & w^2 \\ 1 & w^3 & w^2 & w^1 \end{bmatrix} \begin{bmatrix} f_0(0) \\ f_0(1) \\ f_0(2) \\ f_0(3) \end{bmatrix} \qquad (2.51)$$

The matrix containing 1 and w^ξ terms can be factorized into two matrices as

$$
\begin{bmatrix} f(0) \\ f(2) \\ f(1) \\ f(3) \end{bmatrix} = \begin{bmatrix} 1 & w^0 & 0 & 0 \\ 1 & w^2 & 0 & 0 \\ 0 & 0 & 1 & w^2 \\ 0 & 0 & 1 & w^3 \end{bmatrix} \begin{bmatrix} 1 & 0 & w^0 & 0 \\ 0 & 1 & 0 & w^0 \\ 1 & 0 & w^2 & 0 \\ 0 & 1 & 0 & w^2 \end{bmatrix} \begin{bmatrix} f_0(0) \\ f_0(1) \\ f_0(2) \\ f_0(3) \end{bmatrix} \tag{2.52}
$$

Attention should be paid to the interchange of $f(1)$ and $f(2)$ in the matrix on the left-hand side of the equation. Computing $f(n)$ via Equation (2.52) requires four multiplication and eight addition operations of complex numbers. In contrast, finding the value of the same elements through Equation (2.50) requires 16 complex multiplications and 12 complex additions. Hence, the total number of mathematical operations is reduced significantly with the use of Equation (2.52), instead of Equation (2.50). It is obvious that the reduction in operations becomes more dramatic when N is much greater than 4.

In MATLAB software, the fast Fourier transformation (FFT) has its specific statement. One can simply use one command to obtain the FFT result:

$$
x = fft(y)
$$

This makes Fourier transformation very simple using MATLAB. Here, $fft(\mathbf{x})$ is the discrete Fourier transformation (DFT) of vector $\mathbf{x}$. If the length of $\mathbf{x}$ is a power of 2, a fast radix-2 fast Fourier transform algorithm is utilized. If not, a slower non-power-of 2 algorithm is employed. For matrices, the FFT operation is applied to each column separately.

2.2.3.3. Fourier Transformation as Applied to Smooth Analytical Signals

The major feature of Fourier transformation is that it transforms analytical signals from the time or space domain into the frequency domain. So it is not strange that it can be applied to smooth noisy analytical signals. The reasoning behind is quite simple. In chemical study, noises are usually generated in instrumental measurement and are called white noises that obey the normal distribution of zero mean and equal variance. In general, noises are of high frequency while analytical signals are of low frequency in the time domain. Hence, after transforming the analytical signals into the frequency domain, if one discards the high-frequency part but keeps the low-frequency part, it is possible to eliminate the white noises present in the signals. An example is provided here to illustrate the treatment.

Figure 2.11 gives an example of showing how to use Fourier transformation to smooth a noisy signal. Plot (a) shows the original noisy signal.

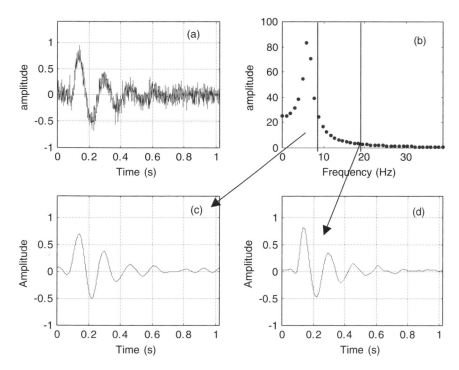

Figure 2.11. Illustration of how Fourier transformation can be applied to smooth an analytical signal: the original noisy signal (a); the dependence of intensity on frequency after Fourier transformation (b); the recovery signal with the cutoff threshold frequency of 10 Hz (c); the recovery signal with a cutoff threshold frequency of 20 Hz (d).

The intensity distribution in the frequency range of 0–40 Hz after Fourier transformation is shown in plot (b). From this plot, it can be seen that the main part of the signal lies in the low-frequency part. The recovery signal obtained with the threshold of cutoff frequency of ∼10 Hz is shown in plot (c). It can be seen that the recovery signal by inverse Fourier transformation is quite smooth with a slight distortion. Plot (d) shows the recovery signal with a cutoff threshold frequency of ∼20 Hz. From the plot, we can see that the recovery signal seems to be better than the one shown in plot (c). Thus, the threshold of the cutoff frequency is an important parameter for smoothing in Fourier transformation. In order to make it easier for the readers to understand the smoothing procedure by Fourier transformation, a MATLAB source code is given as follows:

```
>%demonstration of FT applied to smoothing
>x1=[0.001:.001:1.024];
>for k=101:1024
>    x1b(k)=exp(-(k-100)/176)*sin(2*pi*(k-100)/160);
>end
```

```
>x2=[x1b];
>y=fft(x2);
>subplot(221),plot(x1,x2);grid on
>xlabel('Time (s)');ylabel('amplitude')
>axis([0 1.024 -1 1])
>freq=(0:160)/(1024*.001);
>subplot(222),plot(freq, abs(y(1:161)),'.')
>xlabel('Frequency (Hz)');ylabel('amplitude')
>xx1=ifft(y(1:18),1024);
>subplot(223),plot(real(xx1))
>figure(2)
>x2=x2+randn(size(x2))*.1;
>yy=fft(x2);
>subplot(221),plot(x1,x2);grid on
>xlabel('Time(s)');ylabel('amplitude')
>axis([0 1.024 -1 1.4])
>freq=(0:40)/(1024*.001);
>subplot(222),plot(freq, abs(y(1:41)),'.')
>xlabel('Frequency (Hz)');ylabel('amplitude')
>axis([0 40/(1024*.001) 0 100])
>xx=ifft(yy(1:10),1024);
>subplot(223),plot(x1,real(xx)*2);grid on
>axis([0 1.024 -1 1.4])
>xx=ifft(yy(1:20),1024);
>subplot(224),plot(x1,real(xx)*2);grid on
>axis([0 1.024 -1 1.4])
```

2.2.3.4. *Fourier Transformation as Applied to Convolution and Deconvolution*

Convolution and deconvolution calculations are very important in analytical chemistry, especially in electroanalytical chemistry. Application of Fourier transformation for convolution and deconvolution is based mainly on the following convolution law of

$$f(t) * h(t) \Leftrightarrow F(\nu) \cdot H(\nu) \tag{2.53a}$$

where $F(\nu)$ and $H(\nu)$ correspond to the Fourier transformation of $f(t)$ and $h(t)$.

According to the convolution law, we can first apply Fourier transformation to the functions $f(t)$ and $h(t)$ to obtain the corresponding functions $F(\nu)$ and $H(\nu)$ in the frequency domain, and then carry out convolution calculations on functions $F(\nu)$ and $H(\nu)$. Finally, inverse Fourier transformation is employed to obtain the broadened function $g(t)$ (convolution calculation).

The full procedure is listed as follows:

$$f(t) \xrightarrow{\quad FT \quad} F(v)$$

$$h(t) \xrightarrow{\quad FT \quad} H(v)$$

$$F(v) \cdot H(v) = G(v) \tag{2.53b}$$

$$G(v) \xrightarrow{\text{inverse } FT} g(t)$$

Conversely, since $F(v) = G(v)/H(v)$, the following procedure is utilized to obtain the original signal $f(t)$ (deconvolution calculation):

$$g(t) \xrightarrow{\quad FT \quad} G(v)$$

$$h(t) \xrightarrow{\quad FT \quad} H(v)$$

$$F(v) = \frac{G(v)}{H(v)} \tag{2.54}$$

$$F(v) \xrightarrow{\text{inverse } FT} f(t)$$

This algorithm is also called *inverse filtering*. It should be mentioned that the noise disturbance is completely ignored here. In fact, the measured signal is affected not only by the broadening function $g(t)$ arising from the width of the slit of a device but also by the measurement noise $n(t)$ as

$$g(t) = f(t) * h(t) + n(t) \tag{2.55}$$

Thus, the Fourier transformation $G(v)$ of the measured signal $g(t)$ should be expressed as follows after considering the additive property of Fourier transformation

$$G(v) = F(v)H(v) + N(v) \tag{2.56}$$

$$\frac{G(v)}{H(v)} = F(v) + \frac{N(v)}{H(v)} \tag{2.57}$$

or

$$\widehat{F}(v) = F(v) + \frac{N(v)}{H(v)} \tag{2.58}$$

Here $F(v)$ comes from Fourier transformation of an unknown original signal function $f(t)$, while $\widehat{F}(u)$ is its estimation from deconvolution calculation. One must also consider the influence of noise when performing deconvolution using Fourier transformation.

2.3. NUMERICAL DIFFERENTIATION

Numerical differentiation is a common tool used in analytical chemistry for processing one-dimensional signals because part of the useful information is often "hidden" in the usual measurement plot. The combined chromatographic response, for instance, of two poorly resolved components may not indicate the presence of two coeluting components. The chromatographic profile may appear to be a unimodel and may show no "shoulders" at all. Such "hidden" features can be enhanced by taking the derivative of a signal. A derivative is sensitive to the subtle features of its distribution and is therefore effective in detecting important yet subtle details such as shoulders. In this section, we discuss some common methods used for obtaining derivatives of signals.

2.3.1. Simple Difference Method

The direct-difference method is the simplest numerical differentiation method for analytical signals. For a discrete spectrum x_i ($i = 1, \ldots, n$), suppose that w_i ($i = 1, \ldots, n$) are its sampling wavelengths, the direct-difference method can be expressed as follows:

$$y_i = \frac{x_{i+1} - x_i}{w_{i+1} - w_i} \tag{2.59}$$

Here y_i ($i = 1, \ldots, n - 1$) represent the series of derivatives of spectrum x_i. If the sampling wavelength or time intervals are equal, this equation can be rewritten as

$$y_i = x_{i+1} - x_i \qquad (i = 1, \ldots, n - 1) \tag{2.60}$$

In MATLAB, there is such a function named "diff" for computing derivative spectrum just according to the above equation.

The direct-difference method is simple, but its drawback is also obvious. First, the number of points in the derivative spectrum as deduced in this way is less than one compared to the original discrete spectrum. This leads to the whole derivative spectrum moving half a point forward compared to the original one. Thus, the maximum and/or minimum obtained by the direct-difference method is not exactly the one of zero gradient. Hence, this method is good or acceptable for spectrum of high resolution but not for that of low resolution if one wants to obtain the exact maximum and/or minimum point from the original spectrum. Figure 2.12 shows the results obtained from the "diff" function of MATLAB for both high- and low-resolution spectra.

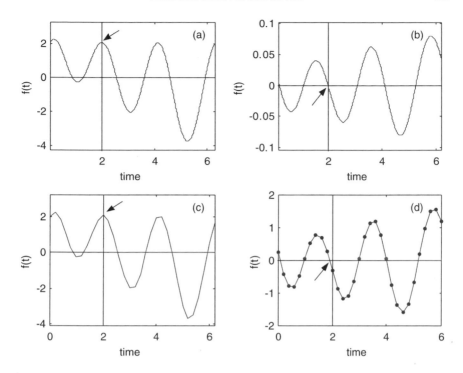

Figure 2.12. The high- (a) and low- (c) resolution spectrum and their derivative spectra (b,d) as obtained by the MATLAB "diff" function.

2.3.2. Moving-Window Polynomial Least-Squares Fitting Method

In Section 2.1.2, we discussed the Savitsky–Golay filter for smoothing, which is essentially a method based on moving-window polynomial least-squares fitting. The filter can also be used for numerical differentiation. As it is based on polynomial fitting, one can directly perform derivation on the polynomial and then obtain the weighting expression for the central point in the moving window to give the derivative spectrum. It is worth noting that the derivative spectrum deduced by this method is better than that obtained by the direct-difference method because there is no point shift in the direct-difference method does.

Every point in a spectrum to be differentiated can be expressed by a polynomial as

$$x_j^{i+j} = a_0 + a_1 j + a_2 j^2 + \cdots + a_k j^k$$

$$(i = 1, \ldots, n)(j = -m, \ldots, 0, \ldots, m) \qquad (2.61)$$

Here the subscripts and superscripts have the same meanings as those in Section 2.1.2. One can directly do derivation on the polynomial. The first derivative is given by

$$\frac{d(x_j^{i+j})}{d(j)} = a_1 + 2a_2 j + \cdots + ka_k j^{k-1} \tag{2.62}$$

With $j = 0$, this equation becomes

$$\frac{d(x_j^{i+j})}{d(j)|_{j=0}} = a_1 \tag{2.63}$$

This means that for the central point ($j = 0$) in the moving window, the weights for the derivative spectrum depends only on the coefficient a_1.

For the second derivative, we have

$$\frac{d^2(x_j^{i+j})}{d(j)} = 2a_2 j + \cdots + (k-1)ka_k j^{k-2} \tag{2.64}$$

With $j = 0$, this becomes

$$\frac{d^2(x_j^{i+j})}{d(j)|_{j=0}} = 2a_2 \tag{2.65}$$

In this way, one can easily obtain the expression of the $(k-1)$-order derivative of the polynomial as

$$\frac{d^k(x_j^{i+j})}{d(j)|_{j=0}} = k!a_k \tag{2.66}$$

Using this expression, we can also find the weights for the derivative spectrum of any order if we know a_i ($i = 1, \ldots, k$). In fact, we can easily determine a_i ($i = 1, \ldots, k$) because of the relationship

$$\mathbf{a} = (\mathbf{M}^t \mathbf{M})\mathbf{M}^t \mathbf{x} \tag{2.67}$$

or

$$\begin{bmatrix} a_0 \\ a_1 \\ \vdots \\ \vdots \\ a_k \end{bmatrix} = (\mathbf{M}^t \mathbf{M})^{-1} \mathbf{M}^t \mathbf{x} \tag{2.68}$$

where $\mathbf{M}$ and $\mathbf{x}$ have the same meanings as those in Section 2.1.2

In this way, Savitsky and Golay collected a_i $(i = 1, \ldots, k)$ with different orders of derivative and with different window sizes in their tables (see Tables 2.3–Table 2.6) for convenience. For instance, we can find the second-derivative spectrum through the use of a window size of $N = 2m + 1 = 9$ and polynomial of fourth or fifth orders through the following expression (see Table 2.5):

$$x_0'''^i = (- 126x_{-4}^{i-4} + 371x_{-3}^{i-3} + 151x_{-2}^{i-2} - 211x_{-1}^{i-1} - 370x_0^i - 211x_1^{i+1}$$
$$+ 151x_2^{i+2} + 371x_3^{i+3} - 126x_4^{i+4})$$

2.4. DATA COMPRESSION

Rapid development in computer technology leads to numerous electronic spectral libraries and databases such as digitized IR, NMR, and mass spectra available in the market. In order to identify the spectrum of an unknown compound from a reference library, data searching is required. Thus, fast and highly accurate library search algorithms are desirable. Before carrying out any search, the spectral library must be constructed first from a set of reference spectra together with additional information such as structure, name, connection data, and molecular mass of each individual compound. In order to reduce the storage space of the spectral data, different compression techniques have been developed.

Generally speaking, the goal of data compression is to represent an information source (e.g., a data file, a speech signal, an image, or a video signal) as accurately as possible using the lowest number of bits. Although many methods have been developed for data compression, most of them are used for compression of images and acoustic signals. Only few of these methods have been adopted in compression of chemical signals.

The principles of data compression methods can be classified into transformation, projection, information encoding, vector quantization, functional approximation, and feature extraction. In this section, data compression methods that are commonly used in chemistry are briefly discussed, including B-spline curve-fitting methods [13–15], Fourier transformation [16–19], and factor analysis [20–22].

2.4.1. Data Compression Based on B-Spline Curve Fitting

The problem of curve fitting is defined as representing a signal by a linear combination of a group of functions. In the case of B-spline curve fitting,

Table 2.3. Weights of Savitsky–Golay Filter for the First Derivative Based on a Cubic/Fourth-Order Polynomial

Points	25	23	21	19	17	15	13	11	9	7	5
-12	30,866										
-11	8,602	3,938									
-10	-8,525	815	84,075								
-9	-20,982	-1,518	10,032	6,936							
-8	-29,236	-3,140	-43,284	68	748						
-7	-33,754	-4,130	-78,176	-4,648	-98	12,922					
-6	-35,003	-4,567	-96,947	-7,481	-643	-4,121	1,133				
-5	-33,450	-4,530	-101,900	-8,700	-930	-14,150	-660	300			
-4	-29,562	-4,098	-95,338	-8,574	-1,002	-18,334	-1,578	-294	86		
-3	-23,806	-3,350	-79,564	-7,372	-902	-17,842	-1,796	-532	-142	22	
-2	-16,649	-2,365	-56,881	-5,363	-673	-13,843	-1,489	-503	-193	-67	1
-1	-8,558	-1,222	-29,592	-2,816	-358	-7,506	-832	-296	-126	-58	-8
0	0	0	0	0	0	0	0	0	0	0	0
1	8,558	1,222	29,592	2,816	358	7,506	832	296	126	58	8
2	16,649	2,365	56,881	5,363	673	13,843	1,489	503	193	67	-1
3	23,806	3,350	79,564	7,372	902	17,842	1,796	532	142	-22	
4	29,562	4,098	95,338	8,574	1,002	18,334	1,578	294	-86		
5	33,450	4,530	101,900	8,700	930	14,150	660	-300			
6	35,003	4,567	96,947	7,481	643	4,121	-1,133				
7	33,754	4,130	78,176	4,648	98	-12,922					
8	29,236	3,140	43,284	-68	-748						
9	20,982	1,518	-10,032	-6,936							
10	8,525	-815	-84,075								
11	-8,602	-3,938									
12	-30,866										
	1,776,060	197,340	3,634,092	255,816	23,256	334,152	24,024	5,148	1,188	252	12

Table 2.4. Weights of Savitsky–Golay Filter for First Derivative Based on a Fifth/Sixth-Order Polynomial

Points	25	23	21	19	17	15	13	11	9	7
−12	−8,322,182									
−11	6,024,183	−400,653								
−10	9,604,353	359,157	−15,033,066							
−9	6,671,883	489,687	16,649,358	−255,102						
−8	544,668	265,164	19,052,988	349,928	−14,404					
−7	−6,301,491	−106,911	6,402,438	322,378	24,661	−78,351				
−6	−12,139,321	−478,349	−10,949,942	9,473	16,679	169,819	−9,647			
−5	−15,896,511	−752,859	−26,040,033	−348,823	−8,671	65,229	27,093	−573		
−4	−17,062,146	−878,634	−24,807,914	−604,484	−32,306	−130,506	−12	2,166	−254	
−3	−15,593,141	−840,937	−35,613,829	−686,099	−43,973	−266,401	−33,511	−1,249	1,381	−1
−2	−11,820,675	−654,687	−28,754,154	−583,549	−40,483	−279,975	−45,741	−3,774	−2,269	9
−1	−6,356,625	−357,045	−15,977,364	−332,684	−23,945	−175,125	−31,380	−3,084	−2,879	−45
0	0	0	0	0	0	0	0	0	0	0
1	6,356,625	357,045	15,977,364	332,684	23,945	175,125	31,380	3,084	2,879	45
2	11,820,675	654,687	28,754,154	583,549	40,483	279,975	45,741	3,774	2,269	−9
3	15,593,141	840,937	35,613,829	686,099	43,973	266,401	33,511	1,249	−1,381	1
4	17,062,146	878,634	34,807,914	604,484	32,306	130,506	12	−2,166	254	
5	15,896,511	752,859	26,040,033	348,823	8,671	−65,229	−27,093	573		
6	12,139,321	478,349	10,949,942	−9,473	−16,679	−169,819	9,647			
7	6,301,491	106,911	−6,402,438	−322,378	−24,661	−169,819				
8	−544,668	−265,164	−19,052,988	−349,928	14,404	78,351				
9	−6,671,883	−489,687	−16,649,358	255,102						
10	−9,604,353	−359,157	15,033,066							
11	−6,024,183	400,653								
12	8,322,182									
	429,214,500	18,747,300	637,408,200	9,806,280	503,880	2,519,400	291,720	17,160	8,580	60

Table 2.5. Weights of Savitsky–Golay Filter for Second Derivative Based on a Quadratic/Cubic Polynomial

Points	25	23	21	19	17	15	13	11	9	7	5
−12	92										
−11	69	77									
−10	48	56	190								
−9	29	37	133	51							
−8	12	20	82	34	40						
−7	−3	5	37	19	25	91					
−6	−16	−8	−2	6	12	52	22				
−5	−27	−19	−35	−5	1	19	11	15			
−4	−36	−28	−62	−14	−8	−8	2	6	28		
−3	−43	−35	−83	−21	−15	−29	−5	−1	7	5	
−2	−48	−40	−98	−26	−20	−44	−10	−6	−8	0	2
−1	−51	−43	−107	−29	−23	−53	−13	−9	−17	−3	−1
0	−52	−44	−110	−30	−24	−56	−14	−10	−20	−4	−2
1	−51	−43	−107	−29	−23	−53	−13	−9	−17	−3	−1
2	−48	−40	−98	−26	−20	−44	−10	−6	−8	0	2
3	−43	−35	−83	−21	−15	−29	−5	−1	7	5	
4	−36	−28	−62	−14	−8	−8	2	6	28		
5	−27	−19	−35	−5	1	19	11	15			
6	−16	−8	−2	6	12	52	22				
7	−3	5	37	19	25	91					
8	12	20	82	34	40						
9	29	37	133	51							
10	48	56	190								
11	69	77									
12	92										
	26,910	17,710	33,649	6,783	3,876	6,188	1,001	429	462	42	7

for example, a signal with m data points, P_i $(1 \cdots m)$, can be represented by a group of B-spline functions

$$\widehat{P}(u) = \sum_{j=1}^{n} c_j N_{j,k}(u) \qquad (2.69)$$

where $N_{j,k}(u)$ represents a k-order B-spline function, c_j is the coefficient of the B-spline function, and u is called a *knot vector*. Therefore, the signal P_i can be represented by n functions and n coefficients.

It is clear that the objective is to find the coefficients so that

$$\sum_{i=1}^{m} (P_i - \widehat{P}_i)^2 \text{ is minimum} \qquad (2.70)$$

Table 2.6. Weights of Savitsky–Golay Filter for Second Derivative Based on a Fourth/Fifth-Order Polynomial

Points	25	23	21	19	17	15	13	11	9	7
−12	−143,198									
−11	10,373	−115,577								
−10	99,385	20,615	−12,597							
−9	137,803	93,993	3,876	−32,028						
−8	138,262	119,510	11,934	15,028	−2,132					
−7	112,067	110,545	13,804	35,148	1,443	−31,031				
−6	69,193	78,903	11,451	36,357	2,691	29,601	−2,211			
−5	18,285	34,815	6,578	25,610	2,405	44,495	2,970	−90		
−4	−33,342	−13,062	626	8,792	1,256	31,856	3,504	174	−126	
−3	−79,703	−57,645	−5,226	−9,282	−207	6,579	1,614	146	371	−13
−2	−116,143	−93,425	−10,061	−24,867	−1,557	−19,751	−971	1	151	67
−1	−139,337	−116,467	−13,224	−35,288	−2,489	−38,859	−3,016	−136	−211	−19
0	−149,290	−124,410	−14,322	−38,940	−2,820	−45,780	−3,780	−190	−370	−70
1	−139,337	−116,467	−13,224	−35,288	−2,489	−38,859	−3,016	−136	−211	−19
2	−116,142	−93,425	−10,061	−24,867	−1,557	−19,751	−971	1	151	67
3	−79,703	−57,645	−5,226	−9,282	−207	6,579	1,614	146	371	−13
4	−33,342	−13,062	626	8,792	1,256	31,856	3,504	174	−126	
5	18,285	34,815	6,578	25,610	2,405	44,495	2,970	−90		
6	69,193	78,903	11,451	36,357	2,691	29,601	−2,211			
7	112,067	110,545	13,804	35,148	1,443	−31,031				
8	138,262	119,510	11,934	15,028	−2,132					
9	137,803	93,993	3,876	−32,028						
10	99,385	20,615	−12,597							
11	10,373	−115,577								
12	−134,198									
	17,168,580	11,248,380	980,628	1,961,256	100,776	1,108,536	58,344	1,716	1,716	132

Such an equation is generally solved using the least-squares method. However, the parameter k must be correctly used because it is closely related to the shape of the signal and it affects the computational speed and the compression ratio. Furthermore, the knot vector is also very important for speeding up the calculation and obtaining better results.

As a development of the B-spline curve-fitting method [15], the B-spline with an order different from that in Equation (2.69) can be replaced by a group of functions with a dilation parameter and a translation parameter:

$$N_{a,b}(t) = N\left(\frac{t-b}{a}\right) \tag{2.71}$$

Equation (2.69) then becomes

$$\widehat{P}(u) = \sum_{j=1}^{n} c_j N_{a_j,b_j}(u) \tag{2.72}$$

With this modification, an analytical signal can be represented by n triplets (a_i, b_i, c_i), which can be obtained by optimization methods, such as genetic algorithms.

For example, curve (a) in Figure 2.13 is a simulated chromatogram comprising four Gaussian peaks with 512 data points. Curve (b) is reconstructed from the compression data as generated from 10 groups of the parameter set (a_i, b_i, c_i) for a second-order B-spline function. The relative root mean square error per point as estimated by the following equation is

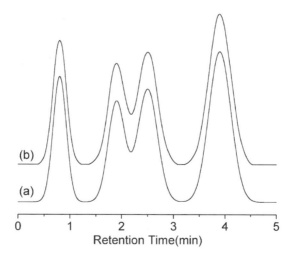

Figure 2.13. A simulated chromatogram (a) and its profile reconstructed from the compressed data (b).

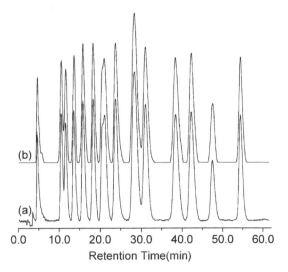

Figure 2.14. An experimental chromatogram (a) and the reconstructed profile using 40 elementary functions (b).

only 0.0012:

$$E = \frac{\left(\sum_i |P_i - \widehat{P}_i|^2\right)^{1/2}}{m\left(\sum_i |P_i|^2\right)^{1/2}} \tag{2.73}$$

Curves (a) in Figures 2.14 and 2.15 are two experimental chromatograms with 900 and 2384 data points, respectively, while curves (b) are profiles

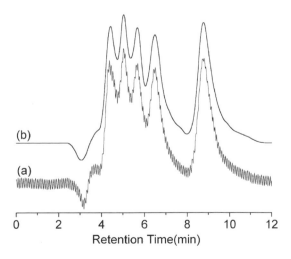

Figure 2.15. An experimental chromatogram (a) and the reconstructed profile using 20 elementary functions (b).

reconstructed by 40 and 20 second-order B-spline functions. The compression ratio is as high as 7.5 : 1 and 39.7 : 1, respectively. From Figure 2.15, it can also be seen that the noise was filtered out during reconstruction. Therefore, such an approach can be used to smooth and compress analytical signals at the same time.

2.4.2. Data Compression Based on Fourier Transformation

As mentioned in Section 2.2, the major feature of Fourier transformation is its transformation of the analytical signals from the time or space domain into the frequency domain. Thus it can also be used for data compression. The principle of compressing the analytical signal by FT is quite simple. It is mainly because the spectra or analytical signals are generally of lower frequency in nature. Consequently, if one keeps only the low-frequency part after transforming the signal into the frequency domain, the signals can be compressed. An example is given here for illustration.

Figure 2.16 gives illustrates how the Fourier transformation is used to compress an infrared spectrum. Curve (a) gives the original simulated infrared spectrum with 3401 data points, while curve (b) shows the result of Fourier transformation on the spectrum from 1 to 1000 frequency points. If only 80 low-frequency points are retained, the recovery signal thus obtained is given as curve (c). It can be seen that the signal reconstructed by using inverse Fourier transformation is quite similar to the original one with a slight distortion. Curve (d) gives the recovery signal with a cutoff threshold frequency of 300 low-frequency points with more data retained compared to the former one following Fourier transformation. It should be noted that the recovery spectrum (Fig. 2.16d) is almost the same as the original one. In this way, an infrared spectrum with 3401 data points can be represented by a data vector containing only 300 low-frequency points with minimal loss of information of the original spectrum. Thus, we can just store these 300 points instead of all the 3401 data of the original spectrum to save the memory space in computer. The procedure of data compression by Fourier transformation is quite similar to that of smoothing the analytical signal. Hence, no source MATLAB code for data compression is provided here.

2.4.3. Data Compression Based on
Principal-Component Analysis

In general, principal-component analysis (PCA) is utilized in chemometrics to solve mainly the problem of calibration and resolution. PCA is used to

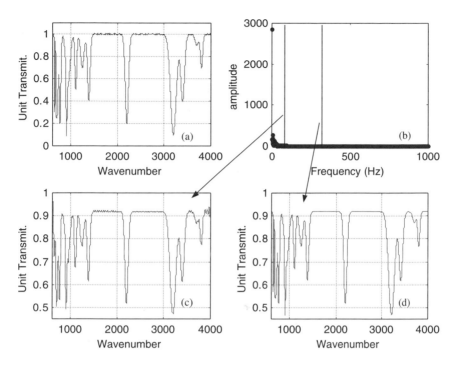

Figure 2.16. Illustration of how the Fourier transformation can be applied to compress a simulated infrared spectrum. The original simulated IR spectrum (a), the dependence of intensity on frequency after Fourier transformation (b), the recovery signal with the cutoff threshold frequency of 80 low-frequency points (c), and the recovery signal with the cutoff threshold frequency of 300 frequency points (d).

decompose the matrix of interest into several independent and orthogonal principal components. The mathematical formula of principal-component analysis can be expressed as follows.

$$\mathbf{X} = \mathbf{U\Theta V}^t = \mathbf{UP}^t = \sum_{i=1}^{A} \mathbf{u}_i \mathbf{p}_i^t \qquad (2.74)$$

where $\mathbf{U\Theta V}^t$ is the singular value decomposition of the matrix $\mathbf{X}$ (see Chapter 5 for more detail). The $\mathbf{u}_i$ and $\mathbf{v}_i$ values ($i = 1, 2, \ldots, A$) are the so-called score and loading vectors, respectively, and are orthogonal with each other. The diagonal matrix $\mathbf{\Theta}$ collects the singular values, which are equal to the square root of the variance (eigenvalues of the covariance matrix $\mathbf{X}^t\mathbf{X}$, i.e., λ_i terms arranged in decreasing order) distributing on every orthogonal principal component axis.

It should be noted that the score and loading vectors are orthogonal to each other and their importance is decided by their corresponding eigenvalues (variance). If datasets of spectra or chromatograms in the matrix

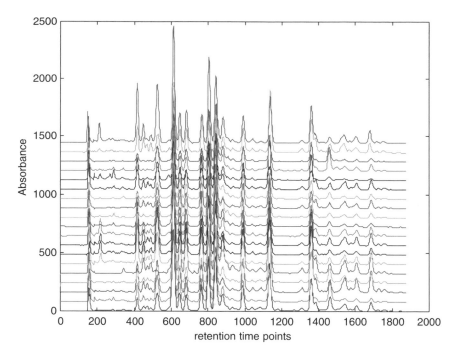

Figure 2.17. Chromatographic fingerprints of 19 *Ginkgo biloba* samples.

form can be represented by a few loading (or score) vectors, data compression by PCA is possible. A simple example is given here to illustrate the data compression procedure of PCA.

Figure 2.17 shows 19 chromatographic fingerprints of *Ginkgo biloba* from different pharmaceutical companies obtained by using high-performance liquid chromatography (HPLC). To store these fingerprints, it seems that all of them have to be included. Could we use PCA to help us to save the memory space of our computer? The answer is "Yes." Let us see how this can be achieved.

The procedure of PCA data compression can be carried out through the following steps:

1. Decompose the data matrix (or dataset) of the analytical signal by PCA to the singular value decomposition [Eq. (2.74)].
2. Find the number of principal components to be retained for reconstructing the original signal later.
3. Store the desirable number of loadings of the largest eigenvalue and the corresponding scores.

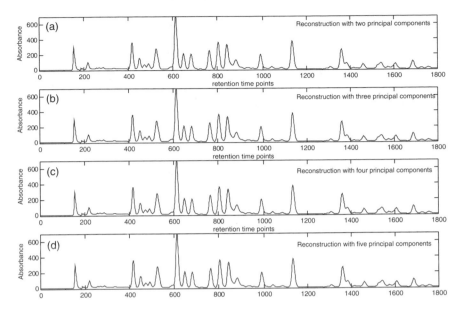

Figure 2.18. Illustration of how PCA can be applied for data compression. From top to bottom: (a) the reconstructed profile of one chromatographic fingerprint by PCA with the first two loadings; (b) the reconstructed profile of one chromatographic fingerprint by PCA with the first three loadings; (c) the reconstructed profile of one chromatographic fingerprint by PCA with the first four loadings; (d) the reconstructed profile of one chromatographic fingerprint by PCA with the first five loadings. Solid blue line—the original fingerprint; dotted red line—the reconstruction fingerprint by PCA.

From this procedure, it can be easily seen that only a few principal loadings and scores are needed to be archived after the PCA treatment. For instance, the variance of the first four principal loadings accounts for 98.32% of the total variance of the 19 samples. Thus, the data to be stored will be the first four principal loadings and the corresponding scores. In comparison to the original dataset, the space for storing the principal loadings and scores for reconstruction is significantly reduced.

Figure 2.18 illustrates the quality of the chromatographic profiles obtained by reconstructing the original analytical signals by different amount of data derived from the principal-component analysis. From these plots, one can see that the chromatographic profile reconstructed by four principal components is good enough because the recovery fingerprint (dotted line) is almost the same as the original one (solid line).

It is worth noting that a new mathematical technique called *wavelet transform* (WT) has been proposed for signal processing in various fields of analytical chemistry owing to its efficiency and speed in data treatment since 1989 [23–25]. Data compression with the use of WT will be discussed in detail in Chapter 5 in this book.

REFERENCES

1. A. Savitsky and M. J. E. Golay, *Anal. Chem.* **36**:1627 (1964).
2. J. Steinier, Y. Termonia, and J. Deltour, *Anal. Chem.* **44**:1906 (1972).
3. P. A. Gorry, *Anal. Chem.* **62**:570 (1990).
4. S. Rutan, *Chemometr. Intell. Lab. Syst.* **6**:191 (1989).
5. S. D. Brown, *Anal. Chim. Acta* **181**:1 (1986).
6. D. L. Massart, B. G. M. Vandeginste, S. N. Deming et al., *Chemometrics: A Textbook*, Elsevier, Amsterdam, 1989.
7. A. G. Marshall and M. B. Comisarow, "Multichannel methods in spectroscopy," in *Transform Techniques in Chemistry*, P. R. Griffiths, ed., Plenum Press, New York, 1978, Chapter 3.
8. J. Zupan, S. Bohance, M. Razinger et al., *Anal. Chim. Acta* **210**:63 (1988).
9. R. W. Ramirez, *The FFT Fundamentals and Concepts*, Prentice-Hall, Englewood Cliffs, NJ, 1985.
10. J. W. Cooper, "Data handling in Fourier transform spectroscopy," in *Transform Techniques in Chemistry*, P. R. Griffiths, ed., Plenum Press, New York, 1978, Chapter 4.
11. E. O. Brigham, *The Fast Fourier Transform*, Prentice-Hall, Englewood Cliffs, NJ, 1974.
12. P. R. Griffiths and J. A. De Haseth, *Fourier Transform Infrared Spectroscopy*, Wiley, New York, 1986.
13. B. K. Alsberg and O. M. Kvalheim, "Compression of Nth-order data arrays by B-splines. 1. Theory," *J. Chemometr.* **7**(1):61–73 (1993).
14. B. K. Alsberg, E. Nodland, and O. M. Kvalheim, "Compression of Nth-order data arrays by B-splines. 2. Application to 2nd-order Ft-Ir spectra," *J. Chemometr.* **8**(2):127–145 (1994).
15. X. G. Shao, F. Yu, H. B. Kou, W. S. Cai, and Z. X. Pan, "A wavelet-based genetic algorithm for compression and de-noising of chromatograms," *Anal. Lett.* **32**(9):1899–1915 (1999).
16. E. F. Crawford and R. D. Larsen, *Anal. Chem.* **49**:508–510 (1977).
17. R. B. Lam, S. J. Foulk, and T. L. Isenhour, *Anal. Chem.* **53**:1679–1684 (1981).
18. P. M. Owens and T. L. Isenhour, *Anal. Chem.* **55**:1548–1553 (1983).
19. F. T. Chau and K. Y. Tam, *Comput. Chem.* **18**:13–20 (1994).
20. C. P. Wang and T. L. Isenhour, *Appl. Spectrosc.* **41**:185–194 (1987).
21. E. R. Malinowski, ed., *Factor Analysis in Chemistry*, 2nd ed., Wiley, New York, 1991.
22. G. Hangac, R. C. Wieboldt, R. B. Lam, and T. L. Isenhour, *Appl. Spectrosc.* **36**:40–47 (1982).
23. X. D. Dai, B. Joseph, and R. L. Motard, in *Wavelet Application in Chemical Engineering*, R. L. Motard and B. Joseph, eds., Kluwer Academic Publishers, 1994, pp. 1–32.
24. F. T. Chau, T. M. Shih, J. B. Gao, and C. K. Chan, *Appl. Spectrosc.* **50**:339–349 (1996).
25. K. M. Leung, F. T. Chau, and J. B. Gao, *Chemometr. Intell. Lab. Syst.* **43**:165–184 (1998).

CHAPTER

3

TWO-DIMENSIONAL SIGNAL PROCESSING TECHNIQUES IN CHEMISTRY

3.1. GENERAL FEATURES OF TWO-DIMENSIONAL DATA

In recent years, numerous "hyphenated instrument" technologies have appeared on the market, such as high-performance liquid chromatography with diode array detection (HPLC-DAD), gas chromatography with mass spectroscopic detection (GC-MS), gas chromatography with infrared spectroscopic detection (GC-IR), high-performance liquid chromatography with mass spectroscopic detection (HPLC-MS), and capillary electrophoresis with diode array detection (CE-DAD). In general, the data produced by the hyphenated instruments are matrices where every row is an object (spectrum) and every column is a variable [the chromatogram at a given wavelength, wavenumber, or m/z (mass/charge) unit] as illustrated in Figure 3.1.

The data obtained by such hyphenated instrumentation in chemistry is generally called *two-dimensional* or *two-way data* and have the following features:

1. The two-dimensional data contain both information of chromatogram and spectra. When a sample is measured by the hyphenated instrument, the data collected can always be arranged as a matrix, say, **X**, where every row is an object (spectrum) and every column is a variable (the chromatogram at a given wavelength, wavenumber, or m/z unit). According to the Lambert–Beer law or similar rules, the matrix can be expressed by the product of two matrices as follows:

$$\mathbf{X} = \mathbf{CS}^t = \sum_{k=1}^{A} \mathbf{c}_k \mathbf{s}_k^t \tag{3.1}$$

here A is the number of absorbing components coexisting in the system, while the $\mathbf{c}_i$ and $\mathbf{s}_k$ ($k = 1, 2, \ldots, A$) values are pure concentration profiles

Chemometrics: From Basics To Wavelet Transform, edited by Foo-tim Chau, Yi-zeng Liang, Junbin Gao, and Xue-guang Shao. Chemical Analysis Series, Vol. 164. ISBN 0-471-20242-8. Copyright © 2004 John Wiley & Sons, Inc.

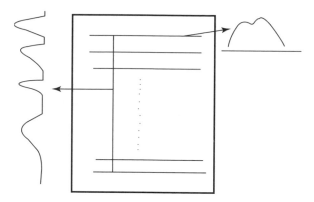

Figure 3.1. Illustration of two-dimensional data from the hyphenated instrument.

and spectra, respectively. The data of Equation (3.1) are called *bilinear two-way data.*

2. The feature of the noise pattern of the two-dimensional data is quite different from that of the one-dimensional data. It should be noted that the two-dimensional (2D) data are contributed from a combination of chromatographic (e.g., LC, GC, HPLC) and a multichannel detectors (e.g., UV, IR, MS). Within a given time interval, a complete spectrum within a specific wavelength range is acquired. Consequently, random errors that occur during chromatographic development will influence the corresponding spectra. Apart from these unavoidable correlated fluctuations, the data will also be contaminated with spurious detector noise, which is sometime correlated between neighboring channels. Moreover, noise that is proportional to the size of the signal is more common than purely additive noise. As a result, the overall noise present in the real data collected from a "spectrochromatograph" will be correlated and, more importantly, the noise should also be heteroscedastic. Thus, pretreatment of 2D data becomes more difficult.

3. The 2D data matrix is usually very large. The 2D matrix can reach capacities of >40 megabytes. Thus, this is really a new challenge in modern analytical chemistry.

In this chapter, we will discuss some basic methods dealing with the two-dimensional data from the so-called hyphenated instruments.

3.2. SOME BASIC CONCEPTS FOR TWO-DIMENSIONAL DATA FROM HYPHENATED INSTRUMENTATION

Before discussing the details of the methods involved, some basic concepts of 2D data from hyphenated instrumentation are mentioned here, to help readers understand these methods.

3.2.1. Chemical Rank and Principal-Component Analysis (PCA)

During chromatographic elution, signals arise and vanish as the chemical constituents appear and disappear at the detector. The variations of the chemical constituents gives rise to corresponding changes in the rank of the two-way data matrix. Thus, rank of the submatrix within a retention-time interval in the raw data matrix is equal to the number of chemical species eluting out simultaneously in the region. This provides a sound mathematical background for solving the problem. Without correct estimation of the number of chemical components (chemical rank) in the system, correct resolution seems to be impossible. However, with measurement errors present in data acquired and other pitfalls in real chemical measurements, correct chemical rank in the two-way data is not trivial at all. Thus, pretreatment of the two-way data and the methodologies for estimating the chemical rankmap sometimes become crucial.

If there were no measurement noises and other pitfalls in measurements, the mathematical rank (the number of independent variables and/or objects in the two-way data) and chemical rank (the number of chemical components in the unknown mixtures) should be the same. Thus, determination of the mathematical rank of a noise-free matrix is trivial. A simple way is to reduce the matrix to the row-echelon form by means of Gaussian elimination and account the number of nonzero rows. However, determination of the chemical rank of a measurement data matrix is a very difficult task because of (1) the presence of measurement noise and their nonassumed distributions, (2) heteroscedasticity of the noise, (3) background and baseline shift arising from the instruments, and (4) collinearity in the measurement data. Thus, in order to avoid these pitfalls, pretreatment of two-way data and local factor analysis becomes very important.

In general, principal-component analysis (or factor analysis) is used in chemometrics to solve the problem of estimating chemical rank in two-way data because it can be used to decompose the matrix into several independent and orthogonal principal components. The number of independent and orthogonal principal components corresponds to the number of the chemical species in the mixture. The mathematical formula of principal component analysis can be expressed as follows:

$$\mathbf{X} = \mathbf{U\Theta V}^t = \mathbf{UP}^t = \sum_{i=1}^{A} \mathbf{u}_i \mathbf{p}_i^t$$

Here $\mathbf{U\Theta V}^t$ represents singular value decomposition of the matrix $\mathbf{X}$ (see Chapter 5 for more detail), while the $\mathbf{u}_i$ and $\mathbf{v}_i$ ($i = 1, 2, \ldots, A$) terms are the so-called score and loading vectors, which are orthogonal with each other. The diagonal matrix Θ collects the singular values, which are equal to the square root of the variance (eigenvalues of the covariance matrix X^tX, i.e., λ_i terms arranged in nonincreasing order) distributed on the every orthogonal principal component axis. Usually, two-way data with measurement noises that are not analyzed by factor analysis can be expressed in the following formula

$$X + E = CS^t + E = U\Theta V^t + E' = UP^t + E' \tag{3.2}$$

or in matrix form as

$$
\begin{bmatrix} x_{11} & x_{12} & \cdots & x_{1N} \\ \cdots & \cdots & \cdots & \cdots \\ \cdots & \cdots & \cdots & \cdots \\ \cdots & \cdots & \cdots & x_{MN} \end{bmatrix} + \begin{bmatrix} e_{11} & e_{12} & \cdots & e_{1N} \\ \cdots & \cdots & \cdots & \cdots \\ \cdots & \cdots & \cdots & \cdots \\ \cdots & \cdots & \cdots & e_{MN} \end{bmatrix}
$$

$$
= \begin{bmatrix} u_{11} & \cdots & u_{1A} \\ u_{21} & \cdots & \cdots \\ \cdots & \cdots & \cdots \\ u_{M1} & \cdots & u_{MA} \end{bmatrix} \cdot \begin{bmatrix} p_{11} & \cdots & \cdots & p_{1N} \\ \cdots & \cdots & \cdots & \cdots \\ p_{A1} & \cdots & \cdots & p_{AN} \end{bmatrix}
$$

$$
+ \begin{bmatrix} u_{1,A+1} & \cdots & u_{1N} \\ u_{2,A+1} & \cdots & \cdots \\ \cdots & \cdots & \cdots \\ u_{M,A+1} & \cdots & u_{MN} \end{bmatrix} \begin{bmatrix} p_{A+1,1} & \cdots & \cdots & p_{A+1,N} \\ \cdots & \cdots & \cdots & \cdots \\ p_{M,1} & \cdots & \cdots & p_{MN} \end{bmatrix} \tag{3.3}
$$

$$
= \begin{bmatrix} u_{11} & \cdots & u_{1A} \\ u_{21} & \cdots & \cdots \\ \cdots & \cdots & \cdots \\ u_{M1} & \cdots & u_{MA} \end{bmatrix} \cdot \begin{bmatrix} p_{11} & \cdots & \cdots & p_{1N} \\ \cdots & \cdots & \cdots & \cdots \\ p_{A1} & \cdots & \cdots & p_{AN} \end{bmatrix}
$$

$$
+ \begin{bmatrix} e'_{11} & \cdots & \cdots & e'_{1N} \\ e'_{21} & \cdots & \cdots & \cdots \\ \cdots & \cdots & \cdots & \cdots \\ e'_{M1} & \cdots & \cdots & e'_{MN} \end{bmatrix}
$$

In Equation (3.2) the matrix E is the measurement noise with the same size as the matrix X and E' is the error matrix, all of which are collectively called *model residuals*. If the measurement noise is really a white noise, which is supposed to be homoscedastic (i.e., with constant variance σ^2) white noise (uncorrelated), then E and E' are almost the same. They will be orthogonal with each other, and also orthogonal with c_i and s_i ($i = 1, 2, \ldots, A$) values and u_i and p_i ($i = 1, 2, \ldots, A$) values. As assumed, the following relationship holds:

$$\tilde{\lambda}_i = \frac{\lambda_i}{(N - \alpha + 1)(M - \alpha + 1)} = k\sigma^2 \tag{3.4}$$

$$[\alpha = A + 1, \dots, \min(N, M)]$$

Therefore, the estimation of the chemical rank can be conducted by comparing the variances between the factors, at which point the variance will equal or be almost equal to a constant (most methods in statistics are based on this comparison). Then, this point can be considered as noise threshold or, by inspecting the pattern of the scores or loadings, the one with the noisy pattern can be regarded as noise. The error theory of principal-component analysis based on the assumption has been thoroughly discussed in the classic book of Malinowski [1]. In some cases, in which the noise behavior is quite closed to the assumed pattern, the chemical rank can be reasonably deduced from the statistical methods discussed in Malinowski's book.

Example 3.1. Figure 3.2 gives a 2D synthetic chromatogram based on the data matrix of three chemical components with some white noises. The spectra and the corresponding chromatograms of these components are shown in Figure 3.3. The results of the factor analysis are shown in Table 3.1.

From the plot as shown in Figure 3.2, the chromatographic profile seems to come from just one component because it seems to have only one

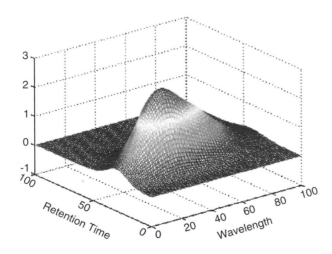

Figure 3.2. Plot of the synthetic 2D data matrix of three chemical components with white noises (see Fig. 3.3 for details).

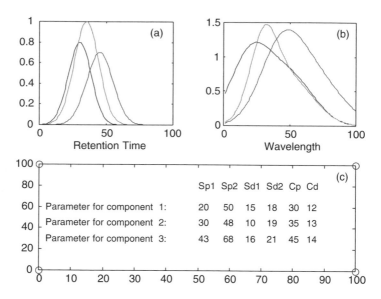

Figure 3.3. The chromatograms and corresponding spectra of the chemical components of the synthetic system shown in Figure 3.2: (a) chromatograms of the three components; (b) spectra of the three components; (c) parameters used for synthesizing the spectra and chromatograms of the components.

peak without any shoulder. However, from the results obtained (Table 3.1), one can see that the first three eigenvalues are significantly larger than the remaining seven eigenvalues, which have relatively very small values. Thus, one can easily conclude that the chromatographic profile as depicted in Figure 3.2 comes from a peak cluster with three components. This example shows that principal-component analysis is really an efficient method

Table 3.1. Eigenvalues of the Simulated Three-Component System

Number	Eigenvalues
1	69.2683
2	8.1592
3	0.9872
4	0.0019
5	0.0019
6	0.0018
7	0.0017
8	0.0017
9	0.0017
10	0.0016

for estimating the number of chemical components in the two-way data matrix.

3.2.2. Zero-Component Regions and Estimation of Noise Level and Background

As discussed before, the task of estimating the number of components becomes very difficult because of the presence of measurement noises and their unknown distributions, heteroscedasticity of the noise and background, and baseline shift from the instruments. To avoid these pitfalls, the determination of zero-component regions from the two-way data as acquired by the hyphenated instrument is of particular significance. The *zero-component region* is defined as the one where no chemical species elute in chromatographic development. Such a region has, by definition, a chemical rank of zero and is of prime importance for establishing the noise level in the data because there is no chemical species present at all within the region. First, if there is no spectral background and chromatographic shift, the signals obtained in the zero-component region are just the noises generated from the instrument within this time interval because the zero-component region plays the role of the analytical blank. With the help of analytical blank information from zero-component regions, an F-test can be established between the two submatrices, in which one is the region to be tested and the other is the region of zero-component. The reader can refer to the paper by Liang et al. [2] for more detail. Furthermore, the zero-component regions may be used to detect the presence of a systematic spectral background and/or baseline offset. This can be done by comparing the zero-component regions before elution of the first species and after the elution of the last species in a peak cluster. Figure 3.4 illustrates a chromatogram with three peak clusters and four zero-component regions.

From Figure 3.4, one can see that no chemical component is eluted in the zero-component regions. Thus, the spectra obtained from these regions should provide the blank information obtained from a spectral detector and should have quite similar spectral backgrounds. We will see later that it is possible for us to estimate the instrumental backgrounds and remove them utilizing the information obtained from the zero-component regions.

From the discussions above, we now focus on how we can correct the background shift in the two-way data. For convenience, raw two-way data in general can be divided into two parts: one originating from the chemical constituents in the analyzed mixture, and the other due to instrumental artifacts, called *spectral background* and *chromatographic baseline shift* here in order to distinguish them from the "random" noise. Thus, raw two-way

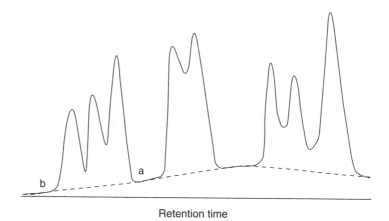

Figure 3.4. A chromatogram with three peak clusters and the zero-component regions before (b) and after (a) the first peak cluster 1. The background is estimated independently for each peak cluster. This means that curved baselines can be locally approximated by straight lines.

data may also be expressed as

$$\mathbf{X} = \mathbf{X}_c + \mathbf{X}_b \quad \text{or} \quad X_{ij} = X_{c,ij} + X_{b,ij}$$
$$(i = 1, \dots, n; \ j = 1, \dots, m) \tag{3.5}$$

where the subscripts c and b denote constituents and background, respectively. Since the most general systematic background for "hyphenated instrument" spectrochromatographic data is a drifting baseline in combination with a spectral background that is approximately constant during the chromatographic run. Such a background of two-way data could be expressed as

$$\mathbf{X}_b = \mathbf{t1}^t + \mathbf{1s}^t \quad \text{or} \quad X_{b,ij} = t_i + s_j \tag{3.6}$$

Here we use vector $\mathbf{t}$ for the baseline shift from chromatography and $\mathbf{s}^t$ for the spectral absorbance vector. The vectors $\mathbf{1}'$ and $\mathbf{1}$ contain only 1s and the dimensions of the two vectors are the number of detector channels n (in wavelength or wavenumbers in spectra) and number of retention timepoints n, respectively. If the random noise is also included in the raw data, the two-way data matrix for the real samples might be simply expressed as follows:

$$\mathbf{X} = \mathbf{CS}^t + \mathbf{t1}^t + \mathbf{1s}^t + \mathbf{E} = \sum_{k=A} \mathbf{c}_k \mathbf{s}_k^t + \mathbf{t1}^t + \mathbf{1s}^t + \mathbf{E} \tag{3.7}$$

All the data pretreatment methods mentioned in the following sections will be based on the discussion in this section.

3.3. DOUBLE-CENTERING TECHNIQUE FOR BACKGROUND CORRECTION

The double-centering technique for background correction is a classic method for dealing with the background of 2D data. In this section, we first examine the effect of combined row–column centering on a 2D matrix where the data are affected by the same chromatographic and spectral background as defined by Equation (3.6). As before, we ignore the random noise in the absorbance matrix. Then, the two-way data matrix can be expressed as

$$\mathbf{X} = \mathbf{CS}^t + \mathbf{1b}^t + \mathbf{s1}^t \quad \text{or} \quad X_{ij} = X_{c,ij} + t_i + s_j \qquad (3.8)$$

Double centering is defined as the operation of subtracting the row and column mean from the matrix $\mathbf{X}$ and adding the grand mean of the matrix:

$$Y_{ij} = X_{ij} - \frac{\sum_j X_{ij}}{m} - \frac{\sum_i X_{ij}}{n} + \frac{\sum_i \sum_j X_{ij}}{nm} \qquad (3.9)$$

Here Y_{ij} is the element at the ith row and the jth column in the data matrix after double centering. The grand mean of $\mathbf{X}$ is easily found to be

$$\frac{\sum_i \sum_j X_{ij}}{nm} = \frac{\sum_i t_i}{n} + \frac{\sum_j s_j}{m} + \frac{\sum_i \sum_j X_{c,ij}}{nm} \qquad (3.10)$$

It should be noted that, in the zero-component regions, $\mathbf{X}_c$ is by definition a zero matrix, and according to Equations (3.9) and (3.10), double centering gives

$$Y_{ij}^0 = (t_i + s_j) - \left(t_i + \frac{\sum_j s_j}{m} \right) - \left(s_j + \frac{\sum_i t_i}{n} + \frac{\sum_i X_{c,ij}}{n} \right)$$
$$+ \frac{\sum_i \sum_j X_{c,ij}}{nm} \qquad (3.11)$$

The superscript 0 is used to indicate the zero-component region. Substitution of Equation (3.10) into Equation (3.11) gives

$$Y_{ij}^0 = \frac{- \sum_i X_{C_{ij}}}{n} + \frac{\sum_i \sum_j X_{C_{ij}}}{nm} \qquad (3.12)$$

From Equation (3.12), one can find that the elements in the double-centered zero-component regions differ only columnwise and have the same values rowwise. Thus the pretreatment process removes the spectral background and chromatographic shift, say, s_j and t_i, respectively. From this perspective, the double-centering method can be used for dealing with these two factors in 2D data. However, one should also be aware that this has created zero-component regions of rank 1. This will lead to difficulties in assessing the level of random noise from these regions which is a key point in the multicomponent system. Furthermore, from Equation (3.12), we can also see that the nonnegativity constraint on the absorbances is violated in the double-centered data.

This procedure is now repeated for regions with chemical rank larger than zero. Starting from Equation (3.9) and double centering, we obtain the following after some manipulation:

$$Y_{ij}^C = X_{C_{ij}} - \frac{\sum_j s_j}{m} - \frac{\sum_j X_{C_{ij}}}{m} - \frac{\sum_i t_i}{n} + \frac{\sum_i X_{C_{ij}}}{n} - \frac{\sum_i X_{C_{ij}}}{nm} \tag{3.13}$$

Inserting Equation (3.10) into Equation (3.13) gives

$$Y_{ij}^C = X_{C_{ij}} + \frac{\sum_i \sum_j X_{C_{ij}}}{nm} - \frac{\sum_j X_{C_{ij}}}{m} - \frac{\sum_i X_{C_{ij}}}{n} \tag{3.14}$$

Here the superscript C denotes the non-zero-component regions. Inspection of Equation (3.14) reveals that double centering also removes spectral and chromatographic background in the regions where chemical components elute. Again, however, the nonnegativity of the intensity for 2D data is destroyed. Furthermore, except for retention-time regions where all chemical species elute simultaneously, the rank is increased by one by the double-centering procedure. It is because of the introduction of a new vector that is a linear combination of all the existing chemical species coeluted. The rank after double centering remains unchanged since the new vector introduced by the double-centering procedure depends linearly on the chemical component spectra.

From the preceding discussion, it can be seen that although double centering may be used to remove background via Equation (3.7), the procedure has some undesired effects, such as the introduction of artificial rank and negative absorbances. Therefore, another procedure is suggested here to correct the spectral background and chromatographic shift based on congruence analysis.

3.4.   CONGRUENCE ANALYSIS AND LEAST-SQUARES FITTING

As mentioned before, the zero-component region using the spectral detector may provide the missing information about the spectral background, because there is no chemical component eluted in this region. Thus, one may detect the spectral and chromatographic baseline offset with the help of this information. Once the spectral and/or chromatographic baseline

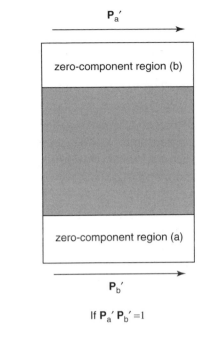

$$\mathbf{P}_a{}'$$

zero-component region (b)

zero-component region (a)

$$\mathbf{P}_b{}'$$

If $\mathbf{P}_a{}' \mathbf{P}_b{}' = 1$

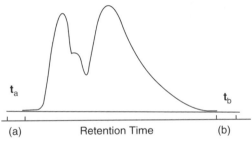

$\mathbf{t}_a$ $\mathbf{t}_b$

(a)   Retention Time   (b)

Figure 3.5. Procedure for correcting the systematic baseline offset and spectral background. The upper part of the figure illustrates the extraction of one principal component in the zero-component regions before (b) elution of the first chemical component start and after (a) the last chemical component has eluted. The loading vectors for the two regions are subsequently compared by means of their scalar product (congruence coefficient). A scalar product with its value close to 1.0 implies the same loading pattern and thus systematic spectral background. The offset vectors, say, $\mathbf{t}_a$ and $\mathbf{t}_b$ in the lower part, in the retention time direction will reveal whether the baseline is drifting. Univariate least-squares fit of the elements of the offset vectors to the retention time and subsequent subtraction corrects the baseline to zero.

offset has been detected in the 2D data, background correction by congruence analysis and least-squares fitting proceeds in the following five steps (see also Fig. 3.5):

1. Calculate the first normalized principal-component loading vector $\mathbf{p}_{1,b}$ for the zero-component region before [Fig. 3.5, segment (b)] elution of

the first chemical component starts and the first normalized principal-component loading vector $\mathbf{p}_{1,a}$ for the zero-component region after [Fig. 3.5, segment (a)] elution of the last chemical component is finished.

2. Compare the two loading vectors by means of their congruence coefficients; that is, calculate the scalar product $\mathbf{p}'_{1,b}\,\mathbf{p}_{1,a}$.

3. If the scalar product in step 2 is close to 1.0, then $\mathbf{p}_{1,b} \approx \mathbf{p}_{1,a}$. This means that the baseline offset can be explained by the same factor (loading vector) during the whole chromatographic elution process. In this case, the offset vector $\mathbf{t}_b$ and $\mathbf{t}_a$ are located for the two zero-component regions.

4. Use the simple univariate least squares to fit a straight line through all the elements of the offset vectors $\mathbf{t}_b$ and $\mathbf{t}_a$ with retention timepoints as independent variables. This procedure provides estimates of the baseline in the whole retention-time region between the two zero-component regions.

5. Collect the estimates in one vector $\mathbf{t}$ and subtract $\mathbf{t1}' + \mathbf{1p}'_{1,b}$ from the data matrix $\mathbf{X}$ to obtain a corrected chromatographic/spectroscopic data matrix. The last step adjusts the baseline to zero level.

This procedure provides a simple way to deal with the spectral and chromatographic baseline offset in two-way data from hyphenated instruments. With the help of this procedure, the spectral and chromatographic baseline offset in the two-way data can be removed with introducing additional artifacts. The entire procedure is also illustrated in Figure 3.5.

3.5. DIFFERENTIATION METHODS FOR TWO-DIMENSIONAL DATA

Two-way data obtained from hyphenated chromatographic instruments are always collected in both retention-time and spectral directions. Thus, the data matrix can be differentiated from both directions. Let matrix $\mathbf{X}$ with n rows and m columns be the 2D array of spectral intensities. The retention time defines the rows, and the spectra define the columns in the matrix. Neglecting the error term, the matrix $\mathbf{X}$ for a mixture of A chemical components can then be expressed as a sum of A bilinear matrices with one for each chemical component:

$$\mathbf{X} = \mathbf{CS}^t = \sum_{k=1}^{A} \mathbf{c}_k \mathbf{s}_k^t \tag{3.15}$$

where $\{\mathbf{c}_k, k = 1, 2, \dots, A\}$ and $\{\mathbf{s}_k^t, k = 1, 2, \dots, A\}$ are the (present unknown) chromatographic and spectral profiles, respectively, of the A

chemical components. The transposition operator t is used to symbolize a row vector as opposed to a column vector.

Assuming that both the retention time and spectral wavelength are continuous parameters, the element $x(t, w)$ of the data matrix $\mathbf{X}$ at time t and wavelength w can be written as

$$x(t, w) = c_1(t)s_1(w) + c_2(t)s_2(w) + \cdots + c_A(t)s_A(w) \tag{3.16}$$

Thus, every element in matrix $\mathbf{X}$ can be described by the functions $\{c_k(t), k = 1, 2, \ldots, A\}$ and $\{s_k(w), k = 1, 2, \ldots, A\}$. Equation (3.16) describes the bilinearity of the matrix $\mathbf{X}$. Furthermore, Equation (3.16) shows that the intensities matrix can be differentiated in either the time or wavelength direction:

$$\frac{dx(t, w)}{dt} = \frac{s_1(w)dc_1(t)}{t} + \frac{s_2(w)dc_2(t)}{dt} + \cdots + \frac{s_A(w)dc_A(t)}{dt} \tag{3.17a}$$

$$\frac{dx(t, w)}{dw} = \frac{c_1(t)ds_1(w)}{w} + \frac{c_2(t)ds_2(w)}{dw} + \cdots + \frac{c_A(t)ds_A(w)}{dt} \tag{3.17b}$$

It should be noted that the derivatives may be expressed in vectorial form as

$$\frac{d\mathbf{x}_i^t(t)}{dt} = \frac{\mathbf{s}_1^t dc_1(t)}{t} + \frac{\mathbf{s}_2^t dc_2(t)}{dt} + \cdots + \frac{\mathbf{s}_A^t dc_A(t)}{dt} \quad (i = 1, 2, \ldots, n) \tag{3.18a}$$

$$\frac{d\mathbf{x}_j(w)}{dw} = \frac{\mathbf{c}_1 ds_1(w)}{w} + \frac{\mathbf{c}_2 ds_2(w)}{dw} + \cdots + \frac{\mathbf{c}_A ds_A(w)}{dt} \quad (i = 1, 2, \ldots, n) \tag{3.18b}$$

Here $\{d\mathbf{x}_i^t(t)/dt \ (i = 1, 2, \ldots, n)\}$ are the row vectors in the derivative matrix $d\mathbf{X}/dt$ obtained by differentiation in the chromatographic (time) direction, while $\{d\mathbf{x}_j(w)/dw \ (j = 1, 2, \ldots, m)\}$ represents the column vectors for the derivative matrix $d\mathbf{X}/dw$ obtained by differentiation in the spectral (wavelength) direction. This means that the two-way data matrix acquired from "hyphenated instruments" can be differentiated from both directions.

Since derivative function has some attractive features, such as enhancing signal resolution and zero value at the maximum of the original function, it has many applications in data analysis for the two-way data. The numerical derivatives of the two-way data can be fulfilled by the technique discussed in Chapter 2. Of course, the wavelet transform technique discussed later on in this book can also be used to complete this task.

3.6. RESOLUTION METHODS FOR TWO-DIMENSIONAL DATA

The ultimate aim for the analysis of 2D data is to simultaneously reveal qualitative and quantitative information about the eluted chemical components as hidden in the chromatographic data that are acquired from hyphenated instrumentation. This topic is discussed in some detail in this section.

In general, when a sample is measured by hyphenated instrumentation, the data can be, as discussed before, collected as a matrix, say $\mathbf{X}$, where every row is an object (spectrum) and every column is a variable (chromatogram at a certain wavelength, wavenumber, or m/z unit). According to the Lambert–Beer law or similar rules, the matrix can be written as

$$\mathbf{X} = \mathbf{CS}^t = \sum_{k=1}^{A} \mathbf{c}_k \mathbf{s}_k^t \qquad (3.19)$$

Here A is the number of absorbing components coexisting in the system, and the $\mathbf{c}_i$ and $\mathbf{s}_i$ ($i = 1, 2, \ldots , A$) terms are the pure concentration profiles and spectra, respectively. The problem to be solved for resolution is that with the measurement matrix at hand, one needs to determine

1. The number of absorbing chemical components A,
2. The spectrum of each chemical component $\mathbf{s}_i$ ($i = 1, 2, \ldots , A$) (qualifications)
3. The concentration profile of each chemical component $\mathbf{c}_i$ ($i = 1, 2, \ldots , A$) (quantification)

One must move from the left-hand side of the equation (LHS; the measurements available from the instrument measurement) to its right-hand side (RHS; the chemical contents from the solution of the problem) with as few artificial assumptions as possible. This type of problem is called "the inverse problem of matrix" in mathematics. In general, it is impossible to obtain a unique solution for the problem. Hence, there may be no mathematical method available at all for solving it uniquely, with a lot of possible solutions from the point of view of mathematics or according to the least-squares criterion. However, the attempt of chemists in trying to directly analyze the two-way data can be traced back in 1960 [3]. This was because the matrix rank has a very good one-to-one correspondence to the number of absorbing chemical components in the system, provided every absorbing chemical component has a different spectrum. Furthermore, the appearance of the *self-modeling curve resolution* (SMCR) [4] based on factor analysis (FA) or principal-component analysis (PCA), rendered the solution of such two-way data problems both possible and attractive. The reason for this is that there are only two well-known constraints from chemical measurements: *i.e.*, non-negative absorbances and concentrations. In particular, the development of "hyphenated" chromatographic instrumentation since the early 1980s—including HPLC-DAD, HPLC-MS, GC-MS, CE-DAD, and GC-IR—made it possible for analysts to directly address the analysis of complex mixture samples. The "hyphenated instruments" possess both the separation ability and quantitative information from chromatographic measurements and plentiful qualitative information from spectral direction. Working with the data obtained using the "hyphenated" instruments, more useful constraints from chemical measurements can be further utilized to narrow the solution space, and a unique solution can sometimes be reached with the help of evolving features and the selective information in the data.

In practice, numerous complex multicomponent systems are employed in analytical chemistry, such as samples from natural products, environmental chemistry, biochemistry, and combinatorial chemistry. It is always the dream of analytical chemists to find a way to determine quickly and accurately how many components there are in these systems, and/or to identify and define these components (qualitative analysis) and/or their amounts (quantitative analysis). The dream may come true now with the development of advanced instruments such as hyphenated chromatographic instruments because vast amounts of useful information about the components can be obtained. However, the volume of data generated from these instruments is humongous, even more than millions of data points from one single experiment! The data are too numerous to be handled by traditional analytical methods, which were developed mainly for handling one-dimensional data. Fortunately, chemometrics that appeared in the early 1970s uses the multivariate analysis technique to solve complex chemical problems. Powerful methodologies have opened new vistas for analytical chemists and have provided meaningful solutions for many complex chemical problems. Combining chemometrics with "hyphenated instruments," it is the right time now in modern analytical chemistry to deal with difficult problems that could not be tackled before.

3.6.1. Local Principal-Component Analysis and Rankmap

As stated before, the size of the data matrix obtained from the hyphenated chromatographic device is vast, sometimes exceeding 40 megabytes. Hence, it is impossible to analyze the matrix simply at one time by PCA at the present stage. On the other hand, collinearity in the matrix may pose another obstacle in chemical rank estimation. Similarity between the component spectra and concentration profiles of the coexisting chemical species makes estimation extremely difficult. The "net analytical signal" for every component will be much smaller because of the overlap between the spectral and concentration profiles of the pure components. When the differences among some component spectra (e.g., isomers or chiral compounds present in a complex sample) are smaller than the level of the noise, how can the mathematical methods be used to recognize these components. Yet, they have different chemical properties! The more components are included in the matrix, the higher the risk of collinearity that one will encounter. If the matrix can be broken down into smaller parts, the risk of collinearity will, of course, be reduced. The first attempt to use the local rank analysis in chemistry to solve the problem was made by Geladi and Wold [5]. They developed an approach called *local rankmapping*, which splits firstly the two-dimensional data set into small parts and then measures their ranks. However, how to split the two-dimensional matrix reasonably is uncertain. The main rationale behind the techniques based on local factor analysis that have been developed in chemometrics is to simply use the separation ability of the chromatography efficiently without losing the information from spectral direction or to figure out the whole rankmap in the chromatographic direction. If the rankmap can be quickly and clearly figured out, the information for resolving efficiently the concentration profiles of all the chemical species in the system is at hand. This is, in our opinion, is the main progress that has been achieved for resolving the two-dimensional data. We will discuss this issue in some detail in the following sections. Here a simple example is utilized to illustrate how 2D data can be rankmapped.

Example 3.2. Figure 3.6 shows a simulated data matrix of three chemical components with some white noise. The simulated spectra and the

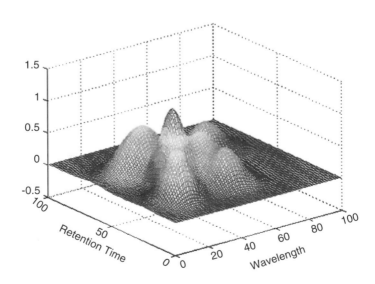

Figure 3.6. A 3D plot of a simulated data matrix of three chemical components with some white noises.

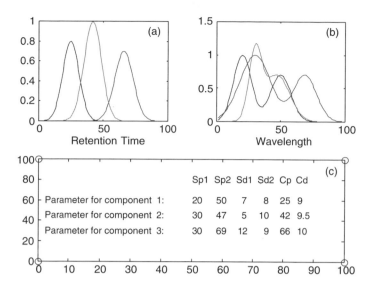

Figure 3.7. The synthetic spectra and the corresponding chromatograms (a,b) of the chemical components together with the parameters used for simulation (c).

corresponding chromatograms of the chemical components together with the parameters used are given in Figure 3.7.

Figure 3.8 illustrates rankmapping of synthetic 2D data in the chromatographic direction. From the figure, one can see that the rankmap clearly indicates the locations of the zero-component regions and the one-component region (selective information), and the mixture regions with rank above 2. With this information at hand, one can easily guess how the chemical components are eluted. If we take the assumption of first come–first disappear, then the elution pattern of the peak cluster can be deduced easily. Now, the problem is how we can obtain such a rankmap. This will be the topic is discussed in the following sections.

3.6.2. Self-Modeling Curve Resolution and Evolving Resolution Methods

As stated before, the development of the so-called self-modeling curve resolution (SMCR) based on factor analysis (FA) or principal-component analysis (PCA) has made resolution of 2D data possible and attractive. The most important feature for SMCR lies in that it has only two well-known constraints from chemical measurements: nonnegative absorbances and concentrations. Thus a number of chemists have done research in the field

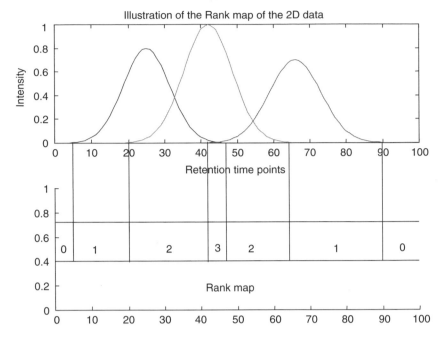

Figure 3.8. A rankmap illustrating the synthetic 2D data in the chromatographic direction.

along this direction. SMCR based only on nonnegative constraints showed that the ambiguities inherent to factor analysis decompositions can be only partly overcome, that is, only a narrow band of possible solutions can be estimated [4].

Current methods available in chemometrics are all based on SMCR. They can be roughly divided into iterative or noniterative types. Examples for methods of the iterative resolution type are iterative target transformation factor analysis (ITTFA) [6], alternative least squares (ALS) [7], the Simplisma method [8], and interative key set factor analysis [9]. The rationale behind this kind of approach lies in introducing further constraints such as unimodality and closure in concentration profiles apart from nonnegative absorbances and concentrations proposed in the original SMCR method for further iteration. If the initial estimate is good enough and the collinearity (overlapping of the signals from both directions) in two-way data is not serious, the methods, in general, can converge to an acceptable solution. However, the ambiguities cannot be eliminated for these methods.

The second type of resolution methods can be regarded as evolutionary. Examples are evolving factor analysis (EFA) [10], window factor analysis (WFA) [11], heuristic evolving latent projections (HELP) [12–14], and more recently, subwindow factor analysis [15]. Their common feature is the use of the informative "windows," such as the selective information regions, zero-concentration regions, and also the regions of $(A - 1)$ components. Thus, the correct estimation of the abovementioned windows based on the local rankmap is crucial in order to obtain the correct resolution results. The resolution condition for "hyphenated" two-way data was comprehensively reviewed by Manne [16]. Three so-called resolution theorems are presented here.

Theorem 3.1. *If all interfering compounds that appear inside the concentration window of a given analyte also appear outside this window, it is possible to calculate the concentration profile of the analyte.*

Theorem 3.2. *If for every interferent the concentration window of the analyte has a subwindow where the interferent is absent, then it is possible to calculate the spectrum of the analyte.*

Theorem 3.3. *For a resolution based on rank information in the chromatographic direction, the condition of Theorems 3.1 and 3.2 are not only sufficient but also necessary.*

As stated before, selectivity is the cornerstone for the resolution of two-way data. But, from Theorems 3.1–3.3, it does not seem necessary to have selective information for complete resolution without ambiguities. Yet, this is not true because the rationale behind these three theorems is that the component that elutes out first should also be the one devolving first in the concentration profiles. Thus, there must be some kind of selective information available in the data. If one cannot mine this out, complete resolution for all the components involved with no ambiguities is impossible. Of course, it is not necessary that every component have its own selective information. But for systems of two components or more, two selective information regions are necessary for complete resolution without ambiguities. The simplest example is the embedded peak with only two components in the chromatogram where only one item of selective information is available. The complete resolution of embedded peaks without additional modeling assumption, to our knowledge, seems impossible for the two-way data so far. Trilinear data may help us obtain the complete resolution desired. The more recent progress for multivariate resolution methodology is based mainly on evolutionary methods. Thus, evolutionary methods are the main focus of this book.

3.6.2.1. Evolving Factor Analysis (EFA)

The first attempt to efficiently use the separation ability in the chromatographic direction for estimating the chemical rankmap might be the evolving
factor analysis developed by Gampp et al. [17], which was proposed pri-
marily to deal with the titration data. This technique was later extended to
the analysis of chromatographic two-way data. Its most attractive feature
is application of the evolving information in the elution direction for titration, chromatography, and other chemical procedures. This opens a new
door for chemometricians to work with two-way data. The methodology
from EFA seems simple. It embraces the spectra to be factor-analyzed
in an incremental fashion and then collects the eigenvalues and plots
them against the retention timepoints (see also Figs. 3.9 and 3.10). This
evolving factor-analyzing procedure can also be conducted in the reverse
direction. Finally, the points of appearance and disappearance of every
chemical component can be determined in this manner. The only assumption involved is that the component that first appears will disappear also
first in an evolving manner.

In order to deal with the peak purity problem for two-way chromatography, Keller and Massart developed a method termed *fixed-size moving-
window factor analysis* (FSMWFA) [18] (see also Figs. 3.11 and 3.12).
Instead of factor-analyzing the data matrix in an incremental fashion, the
method factor-analyzes the spectra in a fixed-sized window and moves
the window along the chromatographic direction. The eigenvalues thus
obtained are also plotted against the retention time. This method has been

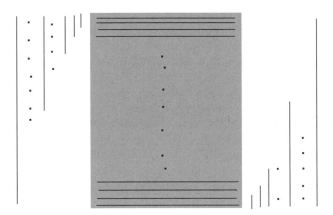

Figure 3.9. Illustration diagram of the evolving factor analysis (EFA) algorithm. The algorithm
can be conducted in both the forward and backward directions. It embraces the spectra to be
factor-analyzed in a stepwise increasing way and then collects the eigenvalues to be plotted
against the retention timepoints.

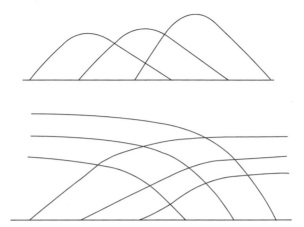

Figure 3.10. The resulting plot of evolving factor analysis of a three-component system. With the help of the results obtained, the points for every chemical component appearing and disappearing can be determined in this way. The only assumption of the method is that the component that first appears will first disappear in an evolving pattern.

successfully applied to detect a minor component with only 0.7% concentration ratio of the major component after correcting the heteroscedastic noise. The procedure in which this method has been utilized to carry out the factor analysis seems to be different from that of EFA, but almost the same information can be extracted. There are two additional advantages of FSMWFA over the original EFA:

1. This method can reduce the calculation time dramatically. The reason is that when the retention timepoints are large enough, which is

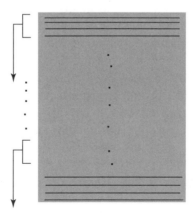

Figure 3.11. Illustration diagram of the fixed-size moving-window evolving factor analysis (FSMWEFA) algorithm. Instead of factor-analyzing the data matrix in a stepwise increasing way, the algorithm is conducted with a moving window. It factor-analyzes the spectra in a fixed-size window and moves the window along the chromatographic direction.

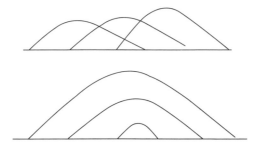

Figure 3.12. The resulting plot of FSMWEFA of a three-component system. With the help of the results obtained, the points for every chemical component appearing and disappearing can be determined more quickly by comparison with the EFA method.

common in two-way data, the EFA will take a long time for the rank estimation, while FSMWEFA takes only a few minutes.

2. The noise level can be built up reasonably since the method is essentially a technique based on local factor analysis. If the noises are correlated slightly, the noise level will be a function of the size of the window analyzed. (see Ref. 4 for further detail).

FSMWEFA is a good tool for local factor analysis, since it collects all the information in the spectral direction. The only thing is that it depends on luck to get the right size of window for the method used to estimate the rank of local regions. Thus, it was extended to a new technique in heuristic evolving latent projections (HELP), called *eigenstructure tracking analysis* (ETA) in order to obtain the whole rankmap in the chromatographic direction [13]. ETA introduces an "evolving size, move window" by starting with a small window first and then increasing the window size in steps by one until the size exceeds the maximum number of overlapping chemical components or is sufficient in the chromatographic regions under investigation. In this way, one deduces not only the maximum resolution with respect to the selective information but also the overlapping information in the retention-time direction. To ensure a correct rankmap, adjustment of moving-window size is sometimes crucial. In fact, with a moving window of increasing size, the sensitivity for detecting chemical signals increases, while with a moving window of decreasing size, the selectivity increases. With this in one's mind, it will be quite helpful for correctly justifying the chemical rank and for obtaining the whole rankmap in the chromatographic direction.

3.6.2.2. Window Factor Analysis (WFA)

The window factor analysis method was developed by Malinowski. It is a self-modeling method for extracting the concentration profiles of individual

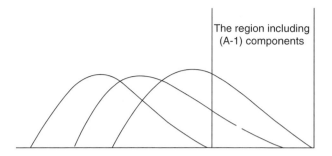

Figure 3.13. Illustration of the major strategy of window factor analysis.

components from evolutionary processes. The region of $(A - 1)$ components, in which the region embracing the analyte is excluded, is used to calculate the concentration profile of the analyte. The major strategy employed in the method is illustrated in Figure 3.13.

The calculation procedure of WFA is quite simple, but the WFA principle is somewhat difficult to understand. For a two-way matrix **X**, it can be decomposed first by PCA in such a way as

$$X = TP^t + E = \sum_{i=1}^{A} t_i p_i^t + E \tag{3.20}$$

Also, according to Lambert–Beer law, the following relation can be formulated:

$$X = CS^t + E = \sum_{i=1}^{A} c_i s_i^t + E \tag{3.21}$$

Here A is the number of chemical components in the system. With the use of matrices **T** and **C**, P^t, and S^t span the same linear space such that

$$c_i = \sum_{i=1}^{A} \beta_{ij} t_j \qquad \text{or} \qquad s_i^t = \sum_{i=1}^{A} \alpha_{ij} p_j^t \tag{3.22}$$

where β_{ij} and α_{ij} are coefficients of the corresponding linear combinations

Suppose that χ is a submatrix that contains only $(A - 1)$ components in matrix **X** as shown in Figure 3.13. Thus, on decomposing this submatrix, the $(A-1)$ orthogonal principal components can be derived via the following procedure:

$$\chi = T^o P^{ot} + E = \sum_{i=1}^{A-1} t_i^o p_i^{ot} + E$$

It should be mentioned that the superscript o is utilized to express these orthogonal vectors in order to distinguish $\mathbf{t}_j^o$ and $\mathbf{t}_j$. As in Equation (3.22), we have

$$\mathbf{c}_i = \sum_{i=1}^{A-1} \eta_{ij} \mathbf{t}_j^o \quad \text{or} \quad \mathbf{s}_i^t = \sum_{i=1}^{A-1} \gamma_{ij} \mathbf{p}_j^{ot} \tag{3.23}$$

Yet, Equations (3.22) and (3.23) differ from each other. In Equation (3.23), the linear space includes only an $(A-1)$-dimensional subspace of $\mathbf{X}$; that is, it has only $(A-1)$ components. In fact, Equation (3.23) can be extended into an A-dimensional space, which can be accomplished by finding a new vector, say, $\mathbf{p}_A^o$, which is orthogonal with all the vectors $\mathbf{p}_j^o (j = 1, \ldots, A-1)$. Then, we have

$$\mathbf{p}_i^t = \sum_{i=1}^{A} \kappa_{ij} \mathbf{p}_j^{ot} \tag{3.24}$$

Owing to the orthogonality of $\mathbf{p}_j^{ot}$, κ_{ij} can be easily determined via

$$\kappa_{ij} = \mathbf{p}_i^t \mathbf{p}_j^o$$

The problem here is how to obtain $\mathbf{p}_A^o$. Summing up all $\mathbf{p}_i^t$ terms in Equation (3.24) gives

$$\sum_{i=1}^{A} \mathbf{p}_i^t = \sum_{i=1}^{A} \left(\sum_{j=1}^{A} \kappa_{ij} \mathbf{p}_j^{ot} \right) = \sum_{j=1}^{A} \left(\sum_{i=1}^{A} \kappa_{ij} \right) \mathbf{p}_j^{ot} \tag{3.25}$$

Equation (3.25) can be rewritten as

$$\sum_{i=1}^{A} \mathbf{p}_i^t = \sum_{i=1}^{A} \left(\sum_{j=1}^{A} \kappa_{ij} \mathbf{p}_j^{ot} \right) = \sum_{j=1}^{A-1} \left(\sum_{i=1}^{A} \kappa_{ij} \right) \mathbf{p}_j^{ot} + \left(\sum_{i=1}^{A} \kappa_{in} \right) \mathbf{p}_A^{ot}$$

and also

$$\left(\sum_{j=1}^{A} \kappa_{in} \right) \mathbf{p}_A^{ot} = \sum_{i=1}^{A} \mathbf{p}_i^t \sum_{j=1}^{A-1} \left(\sum_{i=1}^{A} \kappa_{ij} \right) \mathbf{p}_j^{ot} \tag{3.26}$$

In fact, Equation (3.26) provides a means of computing the vector $\mathbf{p}_A^{ot}$ because all the variables on the RHS of the equation are known or obtainable via PCA, while $(\Sigma \kappa_{in})$ on the LHS of the equation is only a normalized constant for vector $\mathbf{p}_A^{ot}$.

In this way, the orthogonal space $\mathbf{p}_j^{ot} (j = 1, \ldots, A)$ obtained by extension can be expressed linearly by the original linear orthogonal space

$\mathbf{p}_j^t(j = 1, \ldots, A)$; that is, both orthogonal spaces can be related linearly with each other. Thus, the spectra of all the components can be formulated whether by using $\mathbf{p}_j^{ot}(j = 1, \ldots, A)$ or $\mathbf{p}_j^t(j = 1, \ldots, A)$ to give

$$\mathbf{s}_i^t = \sum_{j=1}^{A} \gamma_{ij} \mathbf{p}_j^{ot} \tag{3.27}$$

Inserting Equation (3.27) into the equation $\mathbf{X} = \sum_{i=1}^{A} \mathbf{c}_i \mathbf{s}_i^t$, we obtain

$$\mathbf{X} = \sum_{i=1}^{A} \mathbf{c}_i \left(\sum_{j=1}^{A} \gamma_{ij} \mathbf{p}_j^{ot} \right) = \sum_{i=1}^{A} \sum_{j=1}^{A} \gamma_{ij} \mathbf{c}_i \mathbf{p}_j^{ot}$$

Multiplying $\mathbf{p}_n^o$ on both sides of this last equation leads to

$$\mathbf{X}\mathbf{p}_A^o = \left(\sum_{i=1}^{A} \sum_{j=1}^{A} \gamma_{ij} \mathbf{c}_i \mathbf{p}_j^{ot} \right) \mathbf{p}_A^o$$

As $\mathbf{p}_A^o$ is orthogonal to all the vectors $\mathbf{p}_j^o(j = 1, \ldots, A-1)$, that is, $\mathbf{p}_j^{ot}\mathbf{p}_A^o = 0(j = 1, \ldots, A-1)$, it follows that

$$\mathbf{X}\mathbf{p}_A^o = \left(\sum_{i=1}^{A} \gamma_{in} \mathbf{c}_i \right) \mathbf{p}_A^{ot}\mathbf{p}_A^o \tag{3.28}$$

It should be mentioned that when the subscript i in $\mathbf{s}_i^t$ is less than A, that is, when the spectrum of component i is included in the list of $(A-1)$ components, $\mathbf{s}_i^t$ can be expressed linearly by these $(A-1)$ $\mathbf{p}_j^{ot}$ quantities [see Eq. (3.27)]. Hence, all the γ_{iA} $(i = 1, \ldots, A-1)$ are equal to zeros except for γ_{AA} since only $\mathbf{s}_A^t$ needs $\mathbf{p}_A^o$. In addition, $\mathbf{s}_A^t$ can be normalized to yield

$$\mathbf{X}\mathbf{p}_A^o = \gamma_{AA}\mathbf{c}_A \tag{3.29}$$

This equation indicates that if it is possible to calculate the product of the normalized $\mathbf{p}_A^o$ and matrix $\mathbf{X}$, then both $\mathbf{c}_A$ and a normalized constant γ_{AA} can be determined. In this way, WFA can be apply to obtain the pure concentration profile of the Ath component. This procedure can be repeated for all other pure components to find their concentration profiles and to resolve the mixture system.

From the discussion above, the concrete algorithm for WFA consists of the following steps:

1. Find a region containing only $(A-1)$ components in matrix $\mathbf{X}$. This can be achieved with the help of FWSEFA or EFA.

2. Use Equation (3.26), that is

$$\left(\sum_{i=1}^{A} \kappa_{iA}\right) \mathbf{p}_A^{ot} = \sum_{i=1}^{A} \mathbf{p}_i^t - \sum_{j=1}^{A-1} \left(\sum_{i=1}^{A} \kappa_{ij}\right) \mathbf{p}_j^{ot}$$

to deduce $\mathbf{p}_A^{ot}$;

3. Use Equation (3.28) in the form of

$$\mathbf{Xp}_A^o = \gamma_{AA} \mathbf{c}_A$$

to obtain $\mathbf{c}_A$ and then normalize $\mathbf{c}_A$ afterward.

4. Repeat steps 1–3 until all $\mathbf{c}_i$ $(i = 1, 2, \ldots, A)$ are determined. Then the spectra of the corresponding component can be found using the least-squares technique through

$$\mathbf{S}^t = (\mathbf{C}^t \mathbf{C})^{-1} \mathbf{C}^t \mathbf{X}$$

The WFA procedure has been applied successfully to several datasets.

3.6.2.3. Heuristic Evolving Latent Projections (HELP)

HELP differs from EFA and WFA mainly by its emphasis on using selective information. The spectra of the analytes with selective information can be directly determined through decomposing the selective region. The concentration profile for the same analyte can also be determined by including both selective information and zero-concentration regions for this component in the resolution calculation, which has been termed *full-rank resolution*. Once the spectral and concentration profiles of the components with selective information are determined, the component stripping procedure can then be followed to continue to resolve the other components. This HELP method has been successfully utilized for solving several different real samples.

The HELP technique is based on the use of the ordered nature of "hyphenated" data and the selective regions appearing as straight-line segments in bivariate score and loading plots. Score and loadings plots have been used extensively in multivariate exploratory analysis for a long time, but their significance has been overlooked for rank estimation and resolution. There are at least four advantages in using the latent projection graph (LPG):

1. In the bivariate score plot, a straight-line segment pointing to the origin suggests selective information in the retention-time direction. As for the bivariate loadings plot, a straight-line segment pointing to the origin suggests selective information in the spectral direction. The

concept of "straight line" here is, of course, in the sense of least squares.

2. The evolving information of the appearance and disappearance of the chemical components in the retention-time direction can also be provided in LPG. If one can produce the three-dimensional LPG for the peak cluster with more than three components, the LPG can provide more depicting insight about the data structure.

3. Information enabling the detection of shifts of the chromatographic baseline and instrumental background is also provided in LPG. If there is an offset in the chromatogram, the points will not concentrate at the origin in the plot even if one includes the zero-component regions in the data.

4. LPG is also a very good diagnostic tool to identify the embedded peaks in chromatogram. This information is very important for resolving concentration profiles (see examples in the following section). The LPG works like a microscope to assist one to see the details of the data structure of two-way data.

The HELP method also emphasizes to use the local factor analysis. Using a method called *eigenstructure tracking analysis* (ETA), one can get the rankmap about the exact number of chemical species at every retention timepoint. The unique resolution of a two-dimensional dataset into chromatograms and spectra of the pure chemical constituents is carried out via local full-rank analysis in the HELP method. In general, the full resolution procedure for the HELP method can be described in the following four steps:

1. Confirm the background and correct a drifting baseline.

2. Determine the number of components, the selective region, and the zero-component region of each component through of the evolving latent projective graph and rankmap on the basis of the eigenstructure tracking analysis.

3. With the help of the selective information and the zero-component region, carry out a unique resolution of the two-dimensional data into pure chromatographic profiles and mass or optical spectra by means of local full-rank analysis.

4. Verify the reliability of the resolved result.

To clarify these points, an example is given here to illustrate how the HELP procedure works to resolve a two-component system.

Figure 3.14 shows a two-component chromatographic profile obtained from a GC-MS instrument. Figure 3.14a shows two selective regions

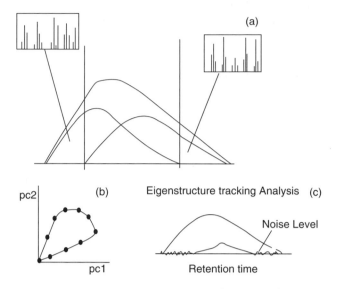

Figure 3.14. A two-component chromatographic profile from a GC-MS instrument: (a) the overlapping chromatographic profile together with the true chromatographic peaks of the two components ; (b) the results obtained by LPG; (c) the results obtained by FSMWFA.

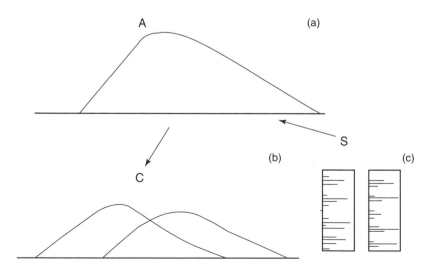

Figure 3.15. The resolved results for the overlapping profile (Fig. 3.14) with two components: (a) the overlapping chromatographic peak; (b) resolved chromatographic profiles; (c) resolved mass spectra.

through which one can easily obtain the pure mass spectra. With the help of LPG and the local factor analysis technique, such as fixed-size moving-window factor analysis (FSMWFA), one can easily locate the selective regions in the chromatographic direction. Figure 3.14a,b shows such results. With the two pure spectra of the two components at hand, the overlapping peaks from the two components can then be resolved easily via least-squares treatment of

$$C = XS(S^tS)^{-1} \qquad (3.30)$$

The resolved results are depicted in Figure 3.15.

REFERENCES

1. E. R. Malinowski, *Factor Analysis in Chemistry*, 2nd ed., Wiley, New York, 1991.
2. Y. Z. Liang, O. M. Kvalheim, and A. Hoskuldsson, "Determination of a multivariate detection limit and local chemical rank by designing a non-parametric test from the zero-component regions," *Chemometr. Intell. Lab. Syst.* **7**:277–290 (1993).
3. R. M. Wallace, "Analysis of absorption spectra of multicomponent systems," *J. Phys. Chem.* **64**:899 (1960).
4. W. E. Lawton and E. A. Sylvestre, "Self modeling curve resolution," *Technometrics* **13**:617–633 (1971).
5. P. Geladi and S. Wold, *Chemometr. Intell. Lab. Syst.* **2**:273–281 (1987).
6. P. J. Gemperline, *J. Chem. Inform. Computer Sci.* **24**:206 (1984).
7. E. J. Karjalainen and U. Karjalainen, *Chemometr. Intell. Lab. Syst.* **14**:423 (1992).
8. W. Windig, "Spectral data files for self-modeling curve resolution with examples using the Simplisma approach," *Chemometr. Intell. Lab. Syst.* **36**:3–16 (1997).
9. K. J. Schostack and E. R. Malinowalski, "Theory of evolutionary factor analysis for resolution of ternary mixtures," *Chemometr. Intell. Lab. Syst.* **8**:121–141 (1990).
10. M. Maeder and A. Zilian, "Evolving factor analysis, a new multivariate technique in chromatography," *Chemometr. Intell. Lab. Syst.* **3**:205–213 (1988).
11. O. M. Kvalheim and Y. Z. Liang, "Heuristic evolving latent projections: —Resolving two-way data multicomponent data. 1. Selectivity, latent projective graphs, datascope, local rank and unique resolution," *Anal. Chem.* **64**:936–945 (1992).
12. Y. Z. Liang, O. M. Kvalheim, H. R. Keller, D. L. Massart, P. Kiechle, and F. Erni, "Heuristic evolving latent projections: —Resolving two-way data multicomponent data. 2. Detection and resolution of minor constituents," *Anal. Chem.* **64**:946–953 (1992).

13. Y. Z. Liang, O. M. Kvalheim, A. Rahmani, and R. Brereton, "Resolution of strongly overlapping two-way multicomponent data by means of heuristic evolving latent projections," *J. Chemometr.* **7**:15–43 (1993).

14. E. R. Malinowski, "Window factor analysis: Theoretical derivation and application to flow injection analysis data," *J. Chemometr.* **6**:29–40 (1992).

15. R. Manne, H. L. Shen, and Y. Z. Liang, "Subwindow factor analysis," *Chemometr. Intell. Lab. Syst.* **45**:171–176 (1999).

16. R. Manne, "On the resolution problem in hyphenated chromatography," *Chemometr. Intell. Lab. Syst.* **27**:89–94 (1996).

17. H. Gampp, M. Maeder, C. J. Meyer, and A. D. Zuberbuhler, *Talanta* **32**:1133 (1985); **33**:943 (1986).

18. H. R. Keller and D. L. Massart, *Anal. Chim. Acta* **246**:279 (1991).

CHAPTER

4

FUNDAMENTALS OF WAVELET TRANSFORM

The objectives of this chapter are to

- Learn about wavelet transform and to learn how to construct a wavelet function.
- Introduce some examples of commonly used wavelet functions such as Haar wavelet, Daubechies wavelet, spline wavelet, and biorthogonal spline wavelet.
- Learn how to implement a wavelet transform and code the algorithm for wavelet transform.
- Learn about the basic concepts of wavelet packet and multivariate wavelet.

In the previous chapters, we introduced several popular analysis methods for digital signal processing in chemistry. One of the main themes of those methods is to decompose a signal into a combination of several (or a countable number of) template signals. For example, fast Fourier transform (FFT) decomposes a signal into a series of sine and cosine waves. As these template signals are very simple to analyze and handle, we can study an original signal and gather information from an original signal by comparing an original signal with those well-understood template signals. However, for different purposes, the templates used should be distinct or purpose-based. The templates in FFT are sine and cosine waves with different frequencies, so the FFT technique can easily tell us the global frequency information contained in a signal. Yet, in some cases we are concerned about the so-called local frequency information. For example, one always wants to locate some spectral peaks corresponding to certain chemicals in spectral analysis. It is desirable to develop new techniques or approaches to provide possible analyzing methods.

From an algorithmic point of view, wavelet analysis offers a harmonious compromise between decomposition and smoothing techniques. Unlike

Chemometrics: From Basics To Wavelet Transform, edited by Foo-tim Chau, Yi-zeng Liang, Junbin Gao, and Xue-guang Shao. Chemical Analysis Series, Vol. 164. ISBN 0-471-20242-8. Copyright © 2004 John Wiley & Sons, Inc.

conventional techniques, wavelet decomposition produces a family of hierarchically organized decompositions. The selection of a suitable level for the hierarchy will depend on the signal and experience. Often, selection of the level is based on a desired lowpass cutoff frequency.

Wavelet transform (WT) has attracted interest in applied mathematics for signal and image processing [7,10,15]. In contrast to some existing popular methods, especially the FFT, this new mathematical technique has been demonstrated to be fast in computation with localization and quick decay properties. Since 1989, WT has been proposed for signal processing in chemical studies owing to its efficiency, large number of basis functions available, and high speed in data treatment. It has been applied successfully in flow injection analysis, high-performance liquid chromatography, infrared spectrometry, mass spectrometry, nuclear magnetic resonance spectrometry, UV–visible spectrometry, and voltammetry for data compression and smoothing. More than a hundred papers have been published on applying WT in chemical studies [1–6,11–13].

4.1. INTRODUCTION TO WAVELET TRANSFORM AND WAVELET PACKET TRANSFORM

As you have noted at the beginning of this chapter, a set or collection of templates should be provided or constructed when we want to analyze an object. Before we introduce the concept of wavelet and multiscale analysis, let us consider the following simple example.

Let us consider measuring a length or distance. First what we need is some standard rules or templates. In the metric system, these templates are units such as kilometer, meter, centimeter, millimeter, and micrometer. Using these templates, we can express (measure or analyze) a length as, for instance, 23 m, 43 cm, and 6 mm. In this procedure, we decompose a length into a (linear) combination of templates; thus we have an idea about the length. Of course, you can choose another template system to measure a length, for example, the imperial unit system. Which system should be chosen is a purpose-based task.

In signal processing, the conventional analyzing tool is Fourier transform. In this technique, a signal to be analyzed is decomposed into a linear combination of standard sine waves at different frequencies, namely, Fourier templates. The benefit from this technique is that we can obtain frequency information in the signal. As the sine wave is a global wave signal, the frequency information obtained will be global. However, sometimes one may want to determine the local frequencies, in which case it will not be easy to extract such information using Fourier transform. We need to create some new techniques to deal with this problem.

In principle, you can choose any signals as your templates. However, to do so is nonsense and does not help solve the problem. Before constructing new templates you should make sure that (1) the templates are suitable to the problem, (2) the template signals are simple to construct and analyze, and (3) the algorithms of analyzing signals under those templates are easily implemented.

WT analysis involves decomposition of a signal function or vector (e.g., a spectrum of a chemical) into a set of approximations and details that are in simpler, fixed building templates at different scales and positions from one "mother" function. The templates are constructed from a "mother" template $\psi(x)$, called "mother wavelet," through scaling $\psi(2^{-j}x)$ and their translates $\psi(2^{-j}x - k)$. In order to understand the meaning of scale, let us consider a simple example. In Figure 4.1, two sine waves, $f(x) = \sin x$ and $g(x) = f(2^3x) = \sin 2^3x$ (i.e., $j = -3$), are shown. The scaled version $g(x)$ of $f(x)$ has higher frequency than does $f(x)$. Thus the concept of frequency can be replaced by the idea of scale. The organizing parameter scale is related to a level j, denoted by 2^{-j}. If we consider $\psi(x)$ as the "0" scale template, then $\psi(2^{-j}x)$ is the jth scale template. Also we call $\psi(2^{-j}x)$ at the resolution 2^j, then the resolution increases as the scale j decreases. The greater the resolution, the smaller and finer are the details that can be accessed.

We should keep in mind the following three questions about the new approach:

1. What kind of template signal is needed? As we want to learn and analyze local information in a signal, the template needed should be compactly

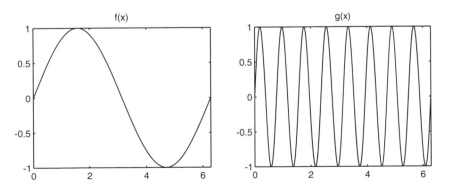

Figure 4.1. Two sine waves at different frequencies in terms of scaling.

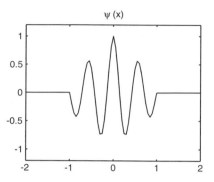

Figure 4.2. The style of a basic template function.

supported.[1] In other words, the template should have an ability to truncate the signal around some special locations. On the other hand, in terms of scaling effect, the template should have an ability to express frequency. Thus the template should be a wave signal. In summary, the basic template $\psi(x)$ would resemble the one shown in Figure 4.2. Thus all the templates $\psi(2^{-j}x)$ will be used to measure or analyze the frequency information of a signal near the time 0. If you want more information about a signal near the time k, just use the translated version of templates $\psi(2^{-j}x - k)$.

2. How many templates do we need? Recall the example of measuring a length. If we didn't have any rulers (templates) finer than the micrometer ruler, we cannot measure anything shorter than 1 μm; that is, the family of templates should be sufficient to extract all information. This requirement is mathematically equivalent to that the family of templates should consist of a basis for the whole signal set.

3. How can we find such a short wavefunction? We should provide a method or theory to describe and construct such wave templates. Many different algorithms were proposed to achieve this goal. In 1989, Mallat [14] introduced the multiresolution signal decomposition (MRSD) algorithm, which was adopted by Daubechies to construct families of compactly supported analyzing template signals. The approach MRSD provides a general method for constructing orthogonal wavelet basis and leads to implementation of the fast wavelet transform (FWT) algorithm.

In the following sections we will consider this theme in the discussions. Let us begin with the simplest wavelet: the Haar wavelet.

---

[1] In mathematics, a function $f(x)$ is said to be compactly supported, if $f(x)$ is zero value outside a finite interval $[a,b]$; that is, if for any x not in $[a,b]$, one has $f(x) = 0$.

4.1.1. A Simple Example: Haar Wavelet

From this subsection, we restrict ourselves in one-dimensional (1D) signal. Generally, any signal or physical phenomenon with finite energy can be represented by a square integrable function; that is, we say a signal $f(x)$ is of finite energy if the signal function satisfies the square-integrable condition

$$\int_{-\infty}^{+\infty} f(x)^2 dx < \infty$$

Mathematically, the set or collection of all such signal functions is denoted by $L^2(\mathbb{R})$. Thus $f(x) \in L^2(\mathbb{R})$ means that the signal function $f(x)$ is square integrable.

In order to introduce Mallat's MRSD algorithm, we first consider Haar basis decomposition of signals here. For simplicity, we denote by $\mathbb{Z}$ the set of (positive and negative) integers, that is, $\mathbb{Z} = \{0, \pm 1, \pm 2, \pm 3, \ldots \}$. In 1910, Haar proposed the following simple piecewise constant function

$$\psi(x) = \begin{cases} -1 & \text{if } 0 \le x < \frac{1}{2} \\ 1 & \text{if } \frac{1}{2} \le x < 1 \\ 0 & \text{otherwise} \end{cases} \tag{4.1}$$

The Haar function $\psi(x)$ is shown in Figure 4.3.

For each pair of integers $j, k \in \mathbb{Z}$, Haar constructed a templates at scales 2^j and translation k

$$\psi_{j,k}(x) = \frac{1}{\sqrt{2^j}} \psi\left(\frac{x - 2^j k}{2^j}\right) \tag{4.2}$$

and made a set of those templates as

$$\mathcal{H} = \{\psi_{j,k}(x) \mid j, k = \ldots, -2, -1, 0, 1, 2, \ldots\} \tag{4.3}$$

For any given (positive or negative) integers j and k, the template $\psi_{j,k}(x)$ is a piecewise constant function with value $-(1/\sqrt{2^j})$ at the interval $[2^j k, 2^j k + 2^{j-1})$, $(1/\sqrt{2^j})$ at the interval $[2^j k + 2^{j-1}, 2^j(k+1))$ and 0 outside the interval $[2^j k, 2^j(k+1)]$. j is called the *scaling level*, or simply level, and k is called the *location*.

Example 4.1. Here are several templates at the different scales and locations which are obtained via Equation (4.2):

$$\psi_{0,-2}(x) = \begin{cases} -1 & -2 \le x < -\frac{3}{2} \\ 1 & -\frac{3}{2} \le x < -1 \\ 0 & \text{otherwise} \end{cases} \qquad \psi_{1,1}(x) = \begin{cases} -\frac{1}{\sqrt{2}} & 2 \le x < 3 \\ \frac{1}{\sqrt{2}} & 3 \le x < 4 \\ 0 & \text{otherwise} \end{cases}$$

$$\psi_{-2,1}(x) = \begin{cases} -2 & \frac{1}{4} \le x < \frac{3}{8} \\ 2 & \frac{3}{8} \le x < \frac{1}{2} \\ 0 & \text{otherwise} \end{cases} \qquad \psi_{3,-1}(x) = \begin{cases} -\frac{1}{2\sqrt{2}} & -8 \le x < -4 \\ \frac{1}{2\sqrt{2}} & -4 \le x < 0 \\ 0 & \text{otherwise} \end{cases}$$

Figure 4.3 shows these Haar templates.

Example 4.1 shows that, the larger the level j is, the boarder wave $\psi_{j,k}(x)$, and the smaller the level j, the sharper the wave of $\psi_{j,k}(x)$ is. Hence a smaller level j corresponds to a finer resolution in $\psi_{j,k}(x)$ and, vice verse, a coarser resolution in $\psi_{j,k}(x)$ is associated with a larger level j.

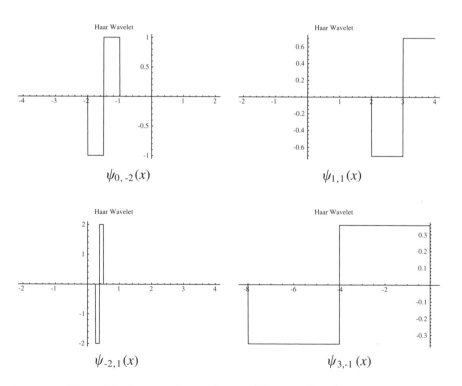

Figure 4.3. Haar wavelet templates at different scales and locations.

First consider the set of finite energy functions $L^2(\mathbb{R})$. Let us definite the inner product of two finite-energy functions $f(x)$ and $g(x)$ from $L^2(\mathbb{R})$ as

$$\langle f(x), g(x) \rangle = \int_{-\infty}^{+\infty} f(x)g(x)dx.$$

As we know, the inner product $\langle f(x), g(x) \rangle$ can be used for measuring the similarity of two signals, $f(x)$ and $g(x)$. The larger the magnitude of inner product is, the greater the similarity is. If the inner product of two functions is 0, then we call these two functions "orthogonal to each other." If function $f(x)$ is orthogonal to a function $g(x)$, then they are said to be *dissimilar*.

The inner product of a function $f(x)$ with itself is called the *energy* of the function. Now, by computing integral, you can easily prove that each Haar template $\psi_{j,k}(x)$ defined in Equation (4.2) has energy 1 and the inner product of any two different Haar templates is 0 (i.e., the different Haar templates are dissimilar). With these two properties, we can say that the Haar collection $\mathcal{H}$ consists of an orthonormal basis of $L^2(\mathbb{R})$, called the *Haar wavelet basis*. Haar further proved that any finite-energy function $f(x)$ can be decomposed into a linear combination form of

$$f(x) = \sum_{j=-\infty}^{+\infty} \sum_{k=-\infty}^{+\infty} \langle f(x), \psi_{j,k}(x) \rangle \psi_{j,k}(x) \tag{4.4}$$

which means that the Haar family is complete or sufficient to represent any finite energy signals.

For the example of measuring a length, Equation (4.4) tells us the length is how many kilometers and how many meters and how many millimeters, and so on. In general, here we can say that the expansion coefficients $\langle f(x), \psi_{j,k}(x) \rangle$ of $\psi_{j,k}(x)$ in Equation (4.4) gives us information on how the signal $f(x)$ of, for instance, a spectrum or a chromatagram is similar to the template signal $\psi_{j,k}(x)$, or we can determine by $\langle f(x), \psi_{j,k}(x) \rangle$ how much of the template $\psi_{j,k}(x)$ a spectrum or a chromatagram has.

As the integral of $\psi_{j,k}(x)$ over $(-\infty, +\infty)$ is zero, each term in summation (4.4) can be interpreted as detailed information on the scale j and the location k. If we put all the detailed information on the scales greater than j together, which is done by truncating finer resolution details from the decomposition (4.4), the new resultant approximation signal

$$f_j(x) = \sum_{m=j+1}^{+\infty} \sum_{k=-\infty}^{+\infty} \langle f(x), \psi_{m,k}(x) \rangle \psi_{m,k}(x) \tag{4.5}$$

can be created. For the Haar wavelet basis $\mathcal{H} = \{\psi_{j,k}(x) | j, k = \ldots, -2, -1, 0, 1, 2, \ldots\}$, $f_j(x)$ is piecewise constant on each interval $[2^j k, 2^j(k+1))$

with width 2^j. When $j \to -\infty$, piecewise function $f_j(x)$ approaches the actual signal $f(x)$. For each integer j, denote by S_j the set of all such piecewise functions $f_j(x)$.

For example, in the case that $j = 0$, S_0 consists of all piecewise functions over intervals $[k, k + 1)$. Define a basic unit step function

$$\phi(x) := \mathbf{1}_{[0,1)} = \begin{cases} 1 & \text{if } 0 \le x < 1 \\ 0 & \text{otherwise} \end{cases} \tag{4.6}$$

Then it is easy to prove that each function $f_0(x)$ in S_0 can be written as a linear combination of translated versions of the basic unit step function $\phi(x)$:

$$f_0(x) = \sum_{k=-\infty}^{\infty} c_k \phi(x - k)$$

In the wavelet community, $\phi(x)$ is called a *scaling function*. In much the same way as we defined the Haar wavelet function $\psi(x)$, we can define a family of dilations/scales and translates for scaling function $\phi(x)$:

$$\left\{ \phi_{j,k}(x) = \frac{1}{\sqrt{2^j}} \phi \left(\frac{x - 2^j k}{2^j} \right) \Big| j, k = \dots, -2, -1, 0, 1, 2, \dots \right\}$$

For each fixed j, any function $f_j(x)$ in S_j can be represented as

$$f_j(x) = \sum_{k=-\infty}^{+\infty} c_{j,k} \phi_{j,k}(x)$$

Example 4.2. Let $f_1(x)$ be a piecewise constant function on half intervals defined as

$$f_1(x) = \begin{cases} 1 & -1 \le x < -\frac{1}{2} \\ 2 & -\frac{1}{2} \le x < 0 \\ 4 & 0 \le x < \frac{1}{2} \\ 5 & \frac{1}{2} \le x < 1 \\ 0 & \text{otherwise} \end{cases}$$

Then we have

$$f_1(x) = \frac{1}{\sqrt{2}} \phi_{-1,-2}(x) + \frac{2}{\sqrt{2}} \phi_{-1,-1}(x) + \frac{4}{\sqrt{2}} \phi_{-1,0}(x) + \frac{5}{\sqrt{2}} \phi_{-1,1}(x)$$

that is, $c_{1,-2} = 1/\sqrt{2}$, $c_{1,-1} = 2/\sqrt{2}$, $c_{1,0} = 4/\sqrt{2}$, $c_{1,1} = 5/\sqrt{2}$ and others $c_{1,k} = 0$.

As $\phi_{0,0}(x)$ is a function in collection S_0 and the functions $\phi_{-1,0}(x)$ and $\phi_{-1,1}$ belong to the collection S_{-1}, note that

$$\phi_{0,0}(x) = \frac{1}{\sqrt{2}}\phi_{-1,0}(x) + \frac{1}{\sqrt{2}}\phi_{-1,1}(x) \tag{4.7}$$

This equation states that the function $\phi_{0,0}(x)$ can be represented as the linear combination of the functions in S_{-1}. On the other hand, each function in S_0 is a form of linear combinations of $\phi_{0,k}$ ($k = \ldots, -2, -1, 0, 1, 2, \ldots$), which can be represented by the functions in S_{-1} [see Eq. (4.7)], then $S_0 \subset S_{-1}$, which means that any function in S_0 can be represented a form of linear combinations of the functions in S_{-1}.

In the example of measuring length, the relation $S_0 \subset S_{-1}$ would be expressed, for instance, in a statement such as 1 km is 1000 m. Generally, one has

$$\cdots \subset S_2 \subset S_1 \subset S_0 \subset S_{-1} \subset S_{-2} \subset \cdots \tag{4.8}$$

and

$$f(x) \in S_0 \Leftrightarrow f\left(\frac{x}{2^j}\right) \in S_j \quad \text{for any } j = \ldots, -2, -1, 0, 1, 2, \ldots. \tag{4.9}$$

Alternatively, we might note that $\psi_{0,0}(x) = \psi(x) \in S_{-1}$ and

$$\psi_{0,0}(x) = -\frac{1}{\sqrt{2}}\phi_{-1,0}(x) + \frac{1}{\sqrt{2}}\phi_{-1,1}(x) \tag{4.10}$$

Example 4.3. Let us consider Example 4.2 again. With the aid of Equations (4.7) and (4.10), we can represent the signal $f_1(x)$ in Example 4.2 as follows:

$$f_1(x) = \frac{1}{\sqrt{2}}\phi_{-1,-2}(x) + \frac{2}{\sqrt{2}}\phi_{-1,-1}(x) + \frac{4}{\sqrt{2}}\phi_{-1,0}(x) + \frac{5}{\sqrt{2}}\phi_{-1,1}(x)$$

$$= \frac{2+1}{2}\phi_{0,-1}(x) + \frac{2-1}{2}\psi_{0,-1}(x) + \frac{5+4}{2}\phi_{0,0}(x) + \frac{5-4}{2}\psi_{0,0}(x)$$

$$= 1.5\phi_{0,-1}(x) + 4.5\phi_{0,0}(x) + 0.5\psi_{0,-1}(x) + 0.5\psi_{0,0}(x)$$

We call this relation *Haar wavelet expansion/representation*. The significance of the Haar wavelet representation can be easily explained by this example. The coefficient 1.5 of $\phi_{0,-1}(x)$ is the mean value of the original signal on the interval $[-1, 0]$ and the coefficient 0.5 of $\psi_{0,-1}(x)$ is the differential or variant of the signal on the same interval. The coefficient 4.5 of $\phi_{0,0}(x)$ is the mean value of the original signal on the interval $[0,1]$ and the coefficient 0.5 of $\psi_{0,0}(x)$ is the differential or variant of the signal on the same interval.

As explained above, we have obtained local average/mean information (like the result given by a moving-average filter) and differential/derivative information (e.g., the result provided by Savitsky–Golay filter) by using Haar wavelet expansion.

4.1.2. Multiresolution Signal Decomposition

Haar's wavelet $\psi(x)$ defined by Equation (4.1) is one simple example of wavelets. It has a simple structure and compact support; however, it is not continuous, as can be seen from Figure 4.3. Are there any other wavelets with better properties, say, continuity with even more smoothness, compact support (i.e., the template function is zero outside a finite interval), and symmetry? How can we find them? Haar's example provides us with a nested approximation structure (4.8) based on Haar wavelet basis decomposition (4.7). In terms of the approximation, the collection S_{j-1} can provide a more accurate approximation to a signal than that by S_j. You can imagine, in the example of measuring a length, that the meter ruler is more accurate than a kilometer ruler and a millimeter ruler is more accurate than a meter ruler, and so forth.

It is very important to note that the Haar wavelet function $\psi(x) \in S_{-1}$ while $\psi(x) \notin S_0$. This fact implies that the Haar wavelet function could be found in the complement of S_0 in S_{-1}. This process can be expressed exactly in terms of a multiresolution signal decomposition (MRSD) first noted by Mallat [14] and used by Daubechies [9,10] to construct a class of new wavelet functions. Although it is very difficult to construct a wavelet function directly, an approximation nested series $\{S_j\}$ with some properties such as (4.8) and (4.9) could be easily provided at sometimes. Then a wavelet function might be constructed from such an approximation nested series.

A multiresolution signal decomposition of $L^2(\mathbb{R})$ is a nested series of closed subspaces $S_j \subset L^2(\mathbb{R})$ (for the example of measuring a length, the following relation means that we have a series of lengths at different scales, " ... lightyear, kilometer, meter, ... ")

$$\{0\} \subset \cdots \subset S_2 \subset S_1 \subset S_0 \subset S_{-1} \subset S_{-2} \subset \cdots \subset L^2(\mathbb{R}) \qquad (4.11)$$

with the following properties[2]

$$\lim_{j \to -\infty} S_j = \overline{\bigcup_{j=-\infty}^{\infty} S_j} = L^2(\mathbb{R}) \tag{4.12}$$

$$\bigcap_{j=-\infty}^{\infty} S_j = \{0\} \tag{4.13}$$

$$f(x) \in S_0 \Leftrightarrow f\left(\frac{x}{2^j}\right) \in S_j \tag{4.14}$$

and there exists a function $\phi(x)$ belonging to $L^2(\mathbb{R})$ whose integer translates

$$\{\phi(x - k) \mid k = \dots, -2, -1, 0, 1, 2, \dots\} \tag{4.15}$$

is an orthonormal basis[3] in $L^2(\mathbb{R})$. We also say that $\phi(x)$ generates a multiresolution signal decomposition $\{S_j\}$.

The term $f_j(x)$ is the representation of f on the scale space S_j and contains all details of $f(x)$ up to finer resolution level j. For example, if the most accurate ruler used is a millimeter ruler, then we obtain a length approximation only down to a millimeter. Property (4.12) says that the signal approximation $f_j(x)$ from S_j converges to an original signal $f(x)$ when $j \to -\infty$ (the precision becomes finer and finer). You can imagine that f_j is the most accurate length to the length $f(x)$ in all measurable lengths S_j under possible rulers; on the other hand, when $j \to +\infty$ (the precision becomes coarser and coarser), Equation (4.13) implies that we lose all the details in the signal $f(x)$. We cannot measure a length with bigger and bigger rulers. Property (4.14) means that S_j is the 2^j scale version of S_0 by changing scale and S_j is spanned by the scaled functions

$$\left\{\phi_{j,k}(x) = \frac{1}{\sqrt{2^j}} \phi\left(\frac{x - 2^j k}{2^j}\right) \,\middle|\, k = \dots, -2, -1, 0, 1, 2, \dots\right\}$$

that is, each element $f(x)$ in S_j (j fixed) can be written in the following form:

$$f(x) = \sum_{k=-\infty}^{+\infty} c_{j,k} \phi_{j,k}(x)$$

For the example of measuring a length, property (4.14) means that a length on a larger scale such as meters can be represented by a length on smaller scales such as millimeters.

---

[2] $A \cup B$ and $A \cap B$ are, respectively, the union and intersection of sets A and B; for instance, let $A = \{1, 2, 3\}$ and $B = \{2, 3, 5, 6\}$, then $A \cup B = \{1, 2, 3, 5, 6\}$ and $A \cap B = \{2, 3\}$. $\bigcup_{j=-\infty}^{\infty} S_j$ is the union of all the collections S_j and $\bigcap_{j=-\infty}^{\infty} S_j$ the intersection of all the collections S_j.

[3] More generally $\{\phi(x - k) \mid k = \dots, -2, -1, 0, 1, 2, \dots\}$ may be a Riesz basis.

The function generating an MRSD is called a *scaling function*. As any scaling function $\phi(x) \in S_0 \subset S_{-1}$, $\phi(x)$ can satisfy a scaling equation, that is, there exists a sequence $H = \{h_k \mid k = \ldots, -2, -1, 0, 1, 2, \ldots\}$ of real numbers with, just like the relation of a length on different scales, we obtain

$$\phi(x) = \sqrt{2} \sum_{k=-\infty}^{+\infty} h_k \phi(2x - k) \qquad (4.16)$$

where the coefficients h_k can be calculated as

$$h_k = \langle \phi(x), \sqrt{2}\phi(2x - k) \rangle \qquad (4.17)$$

This scaling equation relates a scaling function $\phi(x)$ to integer translates of its half-scale versions $\phi_{-2,k}(x)$. The sequence $H = \{h_k \mid k = \ldots, -2, -1, 0, 1, 2, \ldots\}$ in scaling equation might be interpreted as a discrete filter as done in signal processing, called a *scaling filter*.

Example 4.4. Let $\phi(x)$ be the basic unit step function defined by Equation (4.6). It has been shown that the basic unit step yields a multiresolution signal decomposition. From Equation (4.7) the scaling equation [Eq. (4.16)] can be satisfied by taking

$$h_k = \begin{cases} \frac{1}{\sqrt{2}} & \text{if } k = 0, 1 \\ 0 & \text{otherwise} \end{cases}$$

Our aim is to construct a wavelet to describe detailed information in a signal, not just to build a multiresolution signal decomposition. The aim of introducing MRSD is to construct a wavelet function, specifically, the rulers for measuring a length. Hence we are really interested in the approximation error between the approximations of $f(x)$ at the scales j and $j - 1$, which are respectively equal to their orthogonal projections on S_j and S_{j-1}. As S_j is included in S_{j-1}, there is an orthogonal complement of S_j in S_{j-1}:

$$S_{j-1} = S_j \oplus W_j \qquad (4.18)$$

which means that any function in S_{j-1} can be split into two orthogonal parts, one in S_j and the other in W_j. How can we understand this methodology? Consider the problem of measuring length again. Suppose S_{j-1} is all the length information with the accuracy down to the millimeter scale and S_j is the length information on the meter scale. Then we can see that the information in S_{j-1} can be split into the information within S_j and the information exactly on the millimeter scale. Thus our desired ruler on the millimeter scale must lie within the error information between S_{j-1} and S_j.

The subspaces W_j are called *wavelet subspaces*. Similar to the Haar wavelet, we hope that a wavelet basis can be found in these wavelet subspaces. In fact, Mallat [14], Meyer [16], and others proved that for every MRSD there exists a wavelet function $\psi(x)$ whose scaled and translated versions of

$$\left\{ \psi_{j,k}(x) = \frac{1}{\sqrt{2^j}} \psi \left(\frac{x - 2^j k}{2^j} \right) \mid k = \dots, -2, -1, 0, 1, 2, \dots \right\}$$

generate an orthonormal basis of the wavelet subspace W_j for each fixed $j \in \mathbb{Z}$. This is similar to our rulers for measuring a length. Furthermore, the wavelet function can be explicitly constructed from the scaling function $\phi(x)$. The main conclusion can be expressed as follows.

Main Conclusion. Let $\phi(x)$ be an orthogonal scaling function, satisfying the scaling equation (4.16) with a scaling filter $H = \{h_k \mid k = \dots, -2, -1, 0, 1, 2, \dots\}$, and generate an MRSD $\{S_j\}_{j=+\infty}^{-\infty}$. Then the function $\psi(x) \in S_{-1}$, defined by

$$\psi(x) = \sqrt{2} \sum_{k=-\infty}^{+\infty} g_k \phi(2x - k) = \sum_{k=-\infty}^{+\infty} g_k \phi_{-1,k}(x), \qquad (4.19)$$

$$\text{with } g_k = (-1)^{1-k} h_{1-k}, \qquad (4.20)$$

has the following properties:

1. For any fixed j

$$\left\{ \psi_{j,k}(x) = \frac{1}{\sqrt{2^j}} \psi \left(\frac{x - 2^j k}{2^j} \right) \mid k = \dots, -2, -1, 0, 1, 2, \dots \right\}$$

   is an orthonormal basis for W_j
2. All scaled and translated versions $\{\psi_{j,k}(x) \mid j, k = \dots, -2, -1, 0, 1, 2, \dots\}$ provides an orthonormal basis for $L^2(\mathbb{R})$; that is, for any $f(x) \in L^2(\mathbb{R})$, one has

$$f(x) = \sum_{j=-\infty}^{+\infty} \sum_{j=-\infty}^{+\infty} d_{j,k} \psi_{j,k}(x) \qquad (4.21)$$

   where $d_{j,k} = \langle f(x), \psi_{j,k}(x) \rangle$ are called the *wavelet coefficients*.

In the previous equations we called $H = \{h_k \mid k = \dots, -2, -1, 0, 1, 2, \dots\}$ the *scaling filter* and $G = \{g_k \mid k = \dots, -2, -1, 0, 1, 2, \dots\}$ the *wavelet filter*.

We can use two main strategies to construct a wavelet function $\psi(x)$.

Strategy 1

1. Give or choose an MRSD $\{S_j\}_{j=+\infty}^{-\infty}$ associated with a scaling function $\phi(x)$ (sometimes defining an MRSD is much easier than directly constructing a wavelet function).
2. Find the coefficients h_k of scaling filter $H = \{h_k | k = \ldots, -2, -1, 0, 1, 2, \ldots\}$ by Equation (4.17).
3. Construct a wavelet function $\psi(x)$ using Equations (4.19) and (4.20).

Strategy 2

1. Give a scaling filter $H = \{h_k | k = \ldots, -2, -1, 0, 1, 2, \ldots\}$ with some necessary properties and generate a scaling function $\phi(x)$ using Equation (4.16).
2. Construct a wavelet function $\psi(x)$ using Equations (4.19) and (4.20).

4.1.3. Basic Properties of Wavelet Function

The MRSD provides us the techniques for constructing wavelet functions. However, a further question is what kind of wavelet functions are preferred. Most applications of wavelets exploit their ability to efficiently approximate or represent particular classes of signals with few nonzero wavelet coefficients. This applies not only for data compression but also for noise removal in chemical data analysis and fast calculations. The design of wavelet function $\psi(x)$ must therefore be optimized to produce a maximum number of wavelet coefficients $d_{j,k} = \langle f(x), \psi_{j,k}(x) \rangle$ in (4.21) that are close to zero. A signal $f(x)$ has few nonnegligible wavelet coefficients if most of the fine-scale wavelet coefficients are small. This depends mostly on the smoothness of signal $f(x)$ and the properties of wavelet function $\psi(x)$ itself. The following features are very important to a wavelet function with the mentioned optimal property:

Vanishing Moments

A wavelet function $\psi(x)$ has p vanishing moments if

$$\int_{-\infty}^{+\infty} x^k \psi(x) dx = 0 \quad \text{for } 0 \le k < p$$

Why do we introduce the concept of vanishing moments? Simply speaking, if a wavelet function $\psi(x)$ has larger vanishing moments, then the wavelet coefficients $|d_{j,k}|$ of a smooth function $f(x)$ are much smaller on a larger scale j (or at finer resolution).

Compact Support

The compact support of a wavelet function $\psi(x)$ is the maximal interval outside of which wavelet function has zero value. For example, the support of Haar wavelet function $\psi(x)$ is [0, 1) [see Eq. (4.1)]. In general, the smaller the size of such compact, the fewer the high-amplitude wavelet coefficients there are. There exists a relationship between the sizes of support of wavelet function and the corresponding scaling filter. If a scaling filter $\{h_k \mid k = \ldots, -2, -1, 0, 1, 2, \ldots\}$ has only nonzero values for $N_1 \leq k \leq N_2$, then the corresponding wavelet function $\psi(x)$ has a support of size $N_2 - N_1$.

Regularity/Smoothness

The regularity of $\psi(x)$ is a more complicated mathematical concept. It is related to the definition of Hölder continuity and other factors. In this chapter we discuss only continuity and smoothness. For example, the Haar wavelet is an example of discontinous wavelets.

4.2. WAVELET FUNCTION EXAMPLES

In the last section we provided MRSD as a tool to construct some desired wavelet functions. Can any wavelet function be found from a MRSD? The simplest Haar provides us an example that can be constructed by the MRSD generated from a special scaling function: the basic unit step function $\phi(x) = \mathbf{1}_{[0,1)}(x)$. In this section we are about to build more examples and to provide readers with several wavelet functions for their possible applications.

4.2.1. Meyer Wavelet

As mentioned in the rest of Section 4.1.2, one can construct a wavelet function in the Fourier transform. The Meyer wavelet function $\psi(x)$ is the first example of a wavelet function given its Fourier transform, found by French mathematician Y. Meyer in the 1980s. The Meyer wavelet function is a frequency band-limited function whose Fourier transform has a compact support, namely, nonzero only in a finite interval of frequency variable ω. It is constructed by using the second strategy. The resultant scaling function $\phi(x)$ is defined by its Fourier transform. In Meyer's method, only the Fourier transform $\hat{\psi}(\omega)$ of the Meyer wavelet function $\psi(x)$ is given. As the expression is very complicated, we omit it here. Readers can consult Daubechies's book [10] for further details.

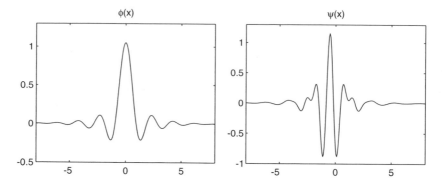

Figure 4.4. Meyer scaling function ϕ and wavelet ψ.

However, we have no explicit formulas for both the Meyer wavelet function $\psi(x)$ and the Meyer scaling function $\phi(x)$, although the Fourier transforms of these two functions have been given. Figure 4.4 displays the corresponding Meyer scaling function $\phi(x)$ and Meyer wavelet function $\psi(x)$. You are encouraged to plot these figures by yourselves with the aid of the MATLAB wavelet toolbox.

Neither the Meyer wavelet function $\psi(x)$ nor the Meyer scaling function $\phi(x)$ has compact support, that is, $\psi(x) \neq 0$ and $\phi(x) \neq 0$ for any x except for some points, but these functions do decrease to 0 when $x \to \infty$ at a very fast speed. Visually they appear as a small wave (see Fig. 4.4), so they can be considered as local waves.

These two functions have better regularity, that is, they are infinitely differentiable, and $\psi(x)$ has an infinite number of vanishing moments; thus, for any integer $m \geq 0$, one has

$$\int_{-\infty}^{+\infty} x^m \psi(x)\,dx = 0$$

4.2.2. B-Spline (Battle–Lemarié) Wavelets

The first strategy is used to construct B-spline wavelet series, assuming a scaling function first. Let $B_m(x)$ be the box spline of degree m, as in Chapter 2. $B_m(x)$ can be computed by convolving the basic unit step function $\mathbf{1}_{[0,1)}$ [see Eq. (4.6)] with itself $(m+1)$ times. It has a compact support centered at 0 or $\frac{1}{2}$, and is a piecewise polynomial function with $(m-1)$ times continuous differentiability. Its Fourier transform is

$$\widehat{B}_m(\omega) = \left(\frac{\sin\frac{\omega}{2}}{\frac{\omega}{2}}\right)^{m+1} e^{-i(v\omega/2)}$$

where v is an indicator whose value is 1 if m is even and 0 if m is odd.

Let S_0 be the set or collection of all the functions consisting of a linear combination of integer translates of $B_m(x)$; thus, for any $f(x)$ in S_0 one has

$$f(x) = \sum_{k \in \mathbb{Z}} c_k B_m(x - k)$$

Since the integer translates $B_m(x - k)$s of B-splines are not orthogonal to each other, we need a procedure for constructing a scaling function $\phi(x)$ in S_0 whose integer translates are orthogonal to each other so that we can generate an MRSD. The procedure can be arrived at in the Fourier domain. After complicated computation, the procedure gives us a scaling function $\phi_m(x)$, which can be represented in terms of its Fourier transform. Similarly, the Fourier transform of a B-spline wavelet $\psi_m(x)$ can be obtained. Neither of these functions is discussed here; instead, properties of both the scaling filter and the wavelet filter are given. Table 4.1 lists h_k corresponding to linear splines $m = 1$ and cubic splines $m = 3$ in which the absolute coefficients less than 10^{-3} are omitted.

The B-spline wavelet function $\psi_m(x)$ has $(m + 1)$ vanishing moments and an exponential decay. Since it is generated by a B-spline of degree m, it is $(m - 1)$ times continuously differentiable. Although the B-spline wavelet has less regularity than the Meyer wavelet, it has faster asymptotic decay; in other words, the B-spline wavelet is more likely to be a local wave than a Meyer wavelet. Figure 4.5 displays the graph of the cubic spline scaling function $\phi_3(x)$ and wavelet $\psi_3(x)$, which is more popular than other B-spline wavelets in actual applications.

Table 4.1. Scaling Coefficients Corresponding to Linear Spline Wavelet and Cubic Spline Wavelet

	k	h_k		k	k_k
$m = 1$	0	0.817645956	$m = 3$	4, −4	0.032080869
	1, −1	0.397296430		5, −5	0.042068328
	2, −2	−0.069101020		6, −6	−0.017176331
	3, −3	−0.051945337		7, −7	−0.017982291
	4, −4	0.016974805		8, −8	0.008685294
	5, −5	0.009990599		9, −9	0.008201477
	6, −6	−0.003883261		10, −10	−0.004353840
	7, −7	−0.002201945		11, −11	−0.003882426
$m = 3$	0	0.766130398		12, −12	0.002186714
	1, −1	0.433923147		13, −13	0.001882120
	2, −2	−0.050201753		14, −14	−0.001103748
	3, −3	−0.110036987			

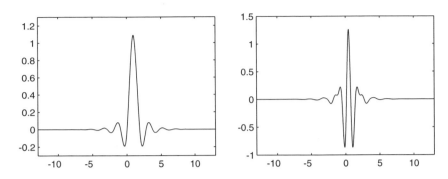

Figure 4.5. Cubic spline scaling function ϕ and wavelet ψ.

4.2.3. Daubechies Wavelets

Daubechies wavelets (see Daubechies wavelets and scaling functions in Fig. 4.6) are commonly used in wavelet analysis. They were developed by mathematician I. Daubechies in 1989. Although Meyer wavelet and spline wavelets are orthogonal, they are not compactly supported; that is, $\psi(x) \neq 0$ for any x except for some points. The Daubechies wavelet is the first example of wavelets with the compactly supported properties. The construction strategy for Daubechies wavelets is to begin with finite-impulse response filters $H = \{h_0, h_1, \ldots, h_{N-1}\}$, which implies that the

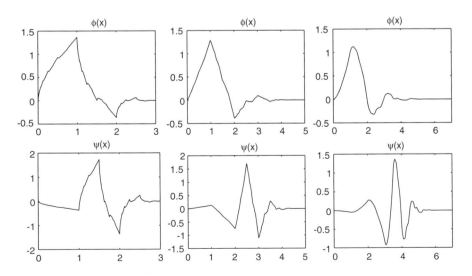

Figure 4.6. Daubechies scaling functions ϕ and wavelets ψ.

transfer function of H is a trigonometric polynomial:

$$H(\omega) = \sum_{k=0}^{N-1} h_k e^{-ik\omega}$$

Daubechies provided a constructive proof of the existence for such a wavelet function. Take $N = 2p$, where p is a given order of vanishing moments; then there exists at least a wavelet $\psi(x)$ with p vanishing moments and support $[-p+1, p]$, and the corresponding scaling function $\phi(x)$ has the support defined by $[0, 2p-1]$. When $p = 1$, one obtains the Haar wavelet. Daubechies gave a computing method for the filter coefficients h_k. Table 4.2 lists all the filter coefficients for $p \leq 7$.

We still have no explicit formulas for the scaling function $\phi(x)$ and the wavelet function $\psi(x)$. The regularity of $\phi(x)$ and $\psi(x)$ is the same since by the scaling Equation (4.19), $\psi(x)$ is a finite linear combination of $\phi_{-1,k}(x)$. However, this regularity is difficult to estimate precisely. In general, the Daubechies wavelet is uniformly smooth. For $p = 2$, the wavelet $\psi(x)$ is only continuous but nondifferentiable (with Lipschitz order 0.55); for $p = 3$, the wavelet $\psi(x)$ is already continuously differentiable.

Daubechies wavelets are very asymmetric. To obtain a symmetric or antisymmetric wavelet, the scaling filter H must be symmetric or antisymmetric with respect to the center of its support. Daubechies has proved that the Haar wavelet is the only one compactly supported classical wavelet whose filter is symmetric or antisymmetric. Daubechies made modifications to her wavelets such that the symmetry of new wavelets can be improved while retaining great simplicity. The resulting wavelets still have a minimum support $[-p+1, p]$ with p vanishing moments, but they are more symmetric. Those wavelets, called *Symmlets*, have many properties similar to those of Daubechies wavelets.

4.2.4. Coiflet Functions

Daubechies wavelets are asymmetric. In actual applications, symmetric templates are sometimes preferred. However, Daubechies proved that no wavelet function could exist with the properties of compact support, continuity, and symmetry occurring simultaneously.

Coiflets are wavelet functions with much more symmetry than Daubechies wavelets. A coiflet function $\psi(x)$ has $2p$ vanishing moments, and the corresponding scaling function $\phi(x)$ has $(2p-1)$ vanishing moments. Both functions have a compact support of length $(6p-1)$.

The coefficients of scaling filters are listed in Table 4.3.

Table 4.2. Daubechies Wavelet Coefficients

	k	h_k		k	k_k
$p = 1$	0	0.707106781	$p = 6$	0	0.111540743
	1	0.707106781		1	0.494623890
$p = 2$	0	0.482962913		2	0.751133908
	1	0.836516304		3	0.315250352
	2	0.224143868		4	−0.226264693
	3	−0.129409523		5	−0.129766868
$p = 3$	0	0.332670553		6	0.097501606
	1	0.806891509		7	0.027522866
	2	0.459877502		8	−0.031582039
	3	−0.135011020		9	0.000553842
	4	−0.085441274		10	0.004777258
	5	0.035226292		11	−0.001077301
$p = 4$	0	0.230377813	$p = 7$	0	0.077852054
	1	0.714846571		1	0.396539319
	2	0.630880768		2	0.729132091
	3	−0.027983769		3	0.469782287
	4	−0.187034812		4	−0.143906004
	5	0.030841382		5	−0.224036185
	6	0.032883012		6	0.071309219
	7	−0.010597402		7	0.080612609
$p = 5$	0	0.160102398		8	−0.038029937
	1	0.603829270		9	−0.016574542
	2	0.724308528		10	0.012550999
	3	0.138428146		11	0.000429578
	4	−0.242294887		12	−0.001801641
	5	−0.032244870		13	0.000353714
	6	0.077571494			
	7	−0.006241490			
	8	−0.012580752			
	9	0.003335725			

For more details, see Daubechies's book [10]. We suggest that you explore coiflets using the MATLAB wavelet toolbox.

4.3. FAST WAVELET ALGORITHM AND PACKET ALGORITHM

In this section we introduce the basic algorithm, derived by Mallat [14] for fast computation of the discrete wavelet transform. In practice, any signals from an equipment, such as chemical spectrum, are discrete. Thus the most useful transform should be in discrete version. The discrete wavelet algorithms can be easily derived from the scaling equations. The wavelet

Table 4.3. The Coefficients of Scaling Filters Corresponding to Coiflets

	k	h_k		k	k_k
$p = 1$	0	−0.015655728	$p = 3$	0	−0.000034600
	1	−0.072732620		1	−0.000070983
	2	0.384864847		2	0.000466217
	3	0.852572020		3	0.001117519
	4	0.337897662		4	−0.002574518
	5	−0.072732612		5	−0.009007976
$p = 2$	0	−0.000720549		6	0.015880545
	1	−0.001823209		7	0.034555028
	2	0.005611435		8	−0.082301927
	3	0.023680172		9	−0.071799822
	4	−0.059434419		10	0.428483476
	5	−0.076488599		11	0.793777223
	6	0.417005184		12	0.405176902
	7	0.812723635		13	−0.061123390
	8	0.386110067		14	−0.065771911
	9	−0.067372555		15	0.023452696
	10	−0.041464937		16	0.007782596
	11	0.016387336		17	−0.003793513

packet algorithm, a more general algorithm, will be also introduced in this section.

4.3.1. Fast Wavelet Transform

Consider a scaling function $\phi(x)$ whose integer translates are orthonormal. Assume that the scaling function $\phi(x)$ generates a MRSD $\{S_j\}_{j=+\infty}^{-\infty}$. For a given discrete signal $\mathbf{c} = \{c_k | k = \ldots, -2, -1, 0, 1, 2, \ldots\}$, let us associate $\mathbf{c}$ with a signal function in S_0:

$$f(x) = \sum_{k=-\infty}^{+\infty} c_k \phi(x - k) \tag{4.22}$$

Mallat developed an algorithm, called *fast wavelet transform* (FWT), to express the signal $f(x)$ of Equation (4.22) in terms of the corresponding wavelet function $\psi(x)$. The algorithm is defined as follows:

$$c_{j,k} = \sum_{m=-\infty}^{+\infty} h_{m-2k} c_{j-1,m}, \tag{4.23}$$

$$d_{j,k} = \sum_{m=-\infty}^{+\infty} g_{m-2k} c_{j-1,m}. \tag{4.24}$$

In the language of signal processing, Equations (4.3) and (4.24) mean that the signals $\mathbf{c}_j = \{c_{j,k} | k = \dots, -2, -1, 0, 1, 2, \dots\}$ and $\mathbf{d}_j = \{d_{j,k} | k = \dots, -2, -1, 0, 1, 2, \dots\}$ are, respectively, the convolutions of $\{c_{j-1,k} | k = \dots, -2, -1, 0, 1, 2, \dots\}$ with the filters $H^* = \{h_{-k} | k = \dots, -2, -1, 0, 1, 2, \dots\} = \{\dots, h_2, h_1, h_0, h_{-1}, h_{-2}, \dots\}$ and $G^* = \{g_{-k} | k = \dots, -2, -1, 0, 1, 2, \dots\} = \{\dots, g_2, g_1, g_0, g_{-1}, g_{-2}, \dots\}$ followed by "down-sampling" of factor 2. Denote still by H^* and G^* such the convolution operators (with downsampling), respectively, then the decomposition algorithms (4.23) and (4.24) can be written as

$$\mathbf{c}_j = H^* \mathbf{c}_{j-1} \qquad (4.25)$$

$$\mathbf{d}_j = G^* \mathbf{c}_{j-1} \qquad (4.26)$$

The whole decomposition process is started from $\mathbf{c}_0 := \mathbf{c}$ until J levels of decomposition where J is a given number of scales. A three-level decomposition process has been shown by Figure 4.7.

After a J-level decomposition process, the initial discrete signal $\mathbf{c}_0$ has been turned into a sequence of newly generated signals $\{\mathbf{c}_J; \mathbf{d}_J; \mathbf{d}_{J-1}; \dots; \mathbf{d}_1\}$.

Example 4.5. In order to see how to implement the decomposition algorithm, consider a special discrete signal $\mathbf{c}_0 = \dots, 0, 0, 1, 2, 2, 1, 0, 0, \dots$ such that $c_{0,-2} = 1$, $c_{0,-1} = 2$, $c_{0,0} = 2$, $c_{0,1} = 1$ and other $c_{0,k} = 0$. Choose a Daubechies wavelet with the filter coefficients:

$$h_0 = \frac{1-\sqrt{3}}{4\sqrt{2}}, \quad h_1 = \frac{3-\sqrt{3}}{4\sqrt{2}}, \quad h_2 = \frac{3+\sqrt{3}}{4\sqrt{2}}, \quad h_3 = \frac{1+\sqrt{3}}{4\sqrt{2}}$$

and from Equation (4.20)

$$g_{-2} = -\frac{1+\sqrt{3}}{4\sqrt{2}}, \quad g_{-1} = \frac{3+\sqrt{3}}{4\sqrt{2}}, \quad g_0 = -\frac{3-\sqrt{3}}{4\sqrt{2}}, \quad g_1 = \frac{1-\sqrt{3}}{4\sqrt{2}}.$$

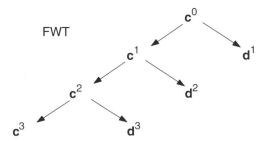

Figure 4.7. The structure of a three-level fast wavelet transform.

At the first-level decomposition, we take $j = 1$ in the algorithms (4.23) and (4.24). By the algorithm we have

$$c_{1,-2} = \sum_{m=-\infty}^{+\infty} h_{m+4} c_{0,m}$$

$$= h_0 c_{0,-4} + h_1 c_{0,-3} + h_2 c_{0,-2} + h_3 c_{0,-1} = \frac{5 + 3\sqrt{3}}{4\sqrt{2}}$$

$$c_{1,-1} = \sum_{m=-\infty}^{+\infty} h_{m+2} c_{0,m}$$

$$= h_0 c_{0,-2} + h_1 c_{0,-1} + h_2 c_{0,0} + h_3 c_{0,1} = \frac{14}{4\sqrt{2}}$$

$$c_{1,0} = \sum_{m=-\infty}^{+\infty} h_m c_{0,m}$$

$$= h_0 c_{0,0} + h_1 c_{0,1} + h_2 c_{0,2} + h_3 c_{0,3} = \frac{5 - 3\sqrt{3}}{4\sqrt{2}}$$

and while for other k, $c_{1,k} = 0$. Hence

$$\mathbf{c}_1 = \ldots, 0, 0, \frac{5 + 3\sqrt{3}}{4\sqrt{2}}, \frac{14}{4\sqrt{2}}, \frac{5 + 3\sqrt{3}}{4\sqrt{2}}, 0, 0, \ldots$$

Similarly, we have

$$d_{1,-1} = \sum_{m=-\infty}^{+\infty} g_{m+2} c_{0,m}$$

$$= g_{-2} c_{0,-4} + g_{-1} c_{0,-3} + g_0 c_{0,-2} + g_1 c_{0,-1} = \frac{-1 - \sqrt{3}}{4\sqrt{2}}$$

$$d_{1,0} = \sum_{m=-\infty}^{+\infty} g_m c_{0,m}$$

$$= g_{-2} c_{0,-2} + g_{-1} c_{0,-1} + g_0 c_{0,0} + g_1 c_{0,1} = \frac{2\sqrt{3}}{4\sqrt{2}}$$

$$d_{1,1} = \sum_{m=-\infty}^{+\infty} g_{m-2} c_{0,m}$$

$$= g_{-2} c_{0,0} + g_{-1} c_{0,1} + g_0 c_{0,2} + g_1 c_{0,3} = \frac{1 - \sqrt{3}}{4\sqrt{2}}$$

and others $d_{1,k} = 0$. Hence

$$\mathbf{d}_1 = \ldots, 0, 0, \quad \frac{-1 - \sqrt{3}}{4\sqrt{2}}, \quad \frac{2\sqrt{3}}{4\sqrt{2}}, \quad \frac{1 - \sqrt{3}}{4\sqrt{2}}, \quad 0, 0, \ldots$$

Similar calculations produce

$$c_{2,-2} = \frac{19 + 14\sqrt{3}}{16}, \quad c_{2,-1} = \frac{22 - 10\sqrt{3}}{16}, \quad c_{2,0} = \frac{7 - 4\sqrt{3}}{16}$$

$$d_{2,-1} = \frac{4 - 9\sqrt{3}}{16}, \quad d_{2,0} = \frac{2 + 10\sqrt{3}}{16}, \quad d_{2,1} = \frac{2 - \sqrt{3}}{16}$$

$$c_{3,-2} = \frac{91 + 73\sqrt{3}}{64\sqrt{2}}, \quad c_{3,-1} = \frac{82 - 62\sqrt{3}}{64\sqrt{2}}, \quad c_{3,0} = \frac{19 - 11\sqrt{3}}{64\sqrt{2}}$$

$$d_{3,-1} = \frac{37 - 55\sqrt{3}}{64\sqrt{2}}, \quad d_{3,0} = -\frac{58 - 22\sqrt{3}}{64\sqrt{2}}, \quad d_{3,1} = \frac{5 - 3\sqrt{3}}{64\sqrt{2}}$$

$$\vdots$$

For example, at three-level decomposition, the final decomposition consists of the following discrete signals:

$$\mathbf{d}_1 = \ldots, 0, 0, \frac{-1 - \sqrt{3}}{4\sqrt{2}}, \frac{2\sqrt{3}}{4\sqrt{2}}, \frac{1 - \sqrt{3}}{4\sqrt{2}}, 0, 0, \ldots$$

$$\mathbf{d}_2 = \ldots, 0, 0, \frac{4 - 9\sqrt{3}}{16}, \frac{2 + 10\sqrt{3}}{16}, \frac{2 - \sqrt{3}}{16}, 0, 0, \ldots$$

$$\mathbf{d}_3 = \ldots, 0, 0, \frac{37 - 55\sqrt{3}}{64\sqrt{2}}, -\frac{58 - 22\sqrt{3}}{64\sqrt{2}}, \frac{5 - 3\sqrt{3}}{64\sqrt{2}}, 0, 0, \ldots$$

$$\mathbf{c}_3 = \ldots, 0, 0, \frac{91 + 73\sqrt{3}}{64\sqrt{2}}, \frac{82 - 62\sqrt{3}}{64\sqrt{2}}, \frac{19 - 11\sqrt{3}}{64\sqrt{2}}, 0, 0, \ldots$$

4.3.2. Inverse Fast Wavelet Transform

Now we turn to the reconstruction problem. Suppose that we have an J-level decomposition $\{\mathbf{c}_J; \mathbf{d}_J; \mathbf{d}_{J-1}; \ldots; \mathbf{d}_1\}$ from a certain discrete signal $\mathbf{c}_0$, where $\mathbf{c}_j = \{c_{j,k} | k = \ldots, -2, -1, 0, 1, 2, \ldots\}$ and $\mathbf{d}_j = \{d_{j,k} | k = \ldots, -2, -1, 0, 1, 2, \ldots\}$. The inverse wavelet transformation can be used to attain the aim of reconstructing a signal $\mathbf{c}_0$; that is, when we are given all the information provided by rulers on different scales, we can reconstruct the information of the whole length. The inverse wavelet is also a successive procedure. In general we can get, at any level j, the *inverse fast wavelet transform* (IFWT)

$$c_{j,k} = \sum_{m=-\infty}^{+\infty} c_{j+1,m} h_{k-2m} + \sum_{m=-\infty}^{+\infty} d_{j+1,m} g_{k-2m} \qquad (4.27)$$

which reconstructs the signal $\mathbf{c}_j$ from $\mathbf{c}_{j+1}$ and $\mathbf{d}_{j+1}$ and can be recursively used to get the original signal $\mathbf{c}_0$ from $j = J$ to $j = 1$.

Let us explain the meaning of reconstruction algorithm (4.27). Define a new signal sequence

$$\hat{c}_l = \begin{cases} c_{j+1,m} & \text{if } l = 2m \\ 0 & \text{if } l = 2m+1 \end{cases} \qquad \text{for } l \in \mathbb{Z}$$

which is created from the sequence $\{c_{j+1,m} | m = \ldots, -2, -1, 0, 1, 2, \ldots\}$ by inserting 0 values between its components. Consider the first part of summation in the right-hand side (RHS) of (4.27). This part can be viewed as the discrete convolutions between the resulted signal $\hat{\mathbf{c}} = \{\hat{c}_l | l = \ldots, -2, -1, 0, 1, 2, \ldots\}$ and the filter $H = \{h_k | k = \ldots, -2, -1, 0, 1, 2, \ldots\}$, that is, following an "upsampling" of factor 2 calculate the convolutions between the upsampled signal and the filter $H = \{h_k | k = \ldots, -2, -1, 0, 1, 2, \ldots\}$. The second part of the summation in (4.27) has a similar explanation. The upsampling-convolution procedures described above are denoted by H and G, respectively; then the algorithm (4.27) is simply written as

$$\mathbf{c}_j = H\mathbf{c}_{j+1} + G\mathbf{d}_{j+1}$$

Example 4.6. From Example 4.5, the two-level decomposed signals are

$$\mathbf{d}_1 = \ldots, 0, 0, \frac{-1-\sqrt{3}}{4\sqrt{2}}, \frac{2\sqrt{3}}{4}, \frac{1-\sqrt{3}}{4\sqrt{2}}, 0, 0, \ldots$$

$$\mathbf{d}_2 = \ldots, 0, 0, \frac{4-9\sqrt{3}}{16}, \frac{2+10\sqrt{3}}{16}, \frac{2-\sqrt{3}}{16}, 0, 0, \ldots$$

$$\mathbf{c}_2 = \ldots, 0, 0, \frac{19+14\sqrt{3}}{16}, \frac{22-10\sqrt{3}}{16}, \frac{7-4\sqrt{3}}{16}, 0, 0, \ldots$$

In order to reconstruct the original signal $\mathbf{c}_0$, we have to get the smooth signal $\mathbf{c}_1$ at the level 1 first.

By the algorithm in Equation (4.27) we have

$$
c_{1,-2} = \sum_{m=-\infty}^{+\infty} c_{2,m} h_{-2-2m} + \sum_{m=-\infty}^{+\infty} d_{2,m} g_{-2-2m}
$$

$$
= c_{2,-2} h_2 + c_{2,-1} h_0 + d_{2,-1} g_0 + d_{2,0} g_{-2}
$$

$$
= \frac{19 + 14\sqrt{3}}{16} \times \frac{3 + \sqrt{3}}{4\sqrt{2}} + \frac{22 - 10\sqrt{3}}{16} \times \frac{1 - \sqrt{3}}{4\sqrt{2}}
$$

$$
+ \frac{4 - 9\sqrt{3}}{16} \times \frac{-3 + \sqrt{3}}{4\sqrt{2}} + \frac{2 + 10\sqrt{3}}{16} \times \frac{-1 - \sqrt{3}}{4\sqrt{2}} = \frac{5 + 3\sqrt{3}}{4\sqrt{2}}
$$

$$
c_{1,-1} = \sum_{m=-\infty}^{+\infty} c_{2,m} h_{-1-2m} + \sum_{m=-\infty}^{+\infty} d_{2,m} g_{-1-2m}
$$

$$
= c_{2,-2} h_3 + c_{2,-1} h_1 + d_{2,-1} g_1 + d_{2,0} g_{-1}
$$

$$
= \frac{19 + 14\sqrt{3}}{16} \times \frac{1 + \sqrt{3}}{4\sqrt{2}} + \frac{22 - 10\sqrt{3}}{16} \times \frac{3 - \sqrt{3}}{4\sqrt{2}}
$$

$$
+ \frac{4 - 9\sqrt{3}}{16} \times \frac{1 - \sqrt{3}}{4\sqrt{2}} + \frac{2 + 10\sqrt{3}}{16} \times \frac{3 + \sqrt{3}}{4\sqrt{2}} = \frac{14}{4\sqrt{2}}
$$

$$
c_{1,0} = \sum_{m=-\infty}^{+\infty} c_{2,m} h_{-2m} + \sum_{m=-\infty}^{+\infty} d_{2,m} g_{-2m}
$$

$$
= c_{2,-1} h_2 + c_{2,0} h_0 + d_{2,0} g_0 + d_{2,1} g_{-2}
$$

$$
= \frac{22 - 10\sqrt{3}}{16} \times \frac{3 + \sqrt{3}}{4\sqrt{2}} + \frac{7 - 4\sqrt{3}}{16} \times \frac{1 - \sqrt{3}}{4\sqrt{2}}
$$

$$
+ \frac{2 + 10\sqrt{3}}{16} \times \frac{-3 + \sqrt{3}}{4\sqrt{2}} + \frac{2 - \sqrt{3}}{16} \times \frac{-1 - \sqrt{3}}{4\sqrt{2}} = \frac{5 - 3\sqrt{3}}{4\sqrt{2}}
$$

Finally, similar calculations will result in the original signal c_0. Take the component $c_{0,0}$ as an example here:

$$
c_{0,0} = \sum_{m=-\infty}^{+\infty} c_{1,m} h_{-2m} + \sum_{m=-\infty}^{+\infty} d_{1,m} g_{-2m}
$$

$$
= c_{1,-1} h_2 + c_{1,0} h_0 + d_{1,0} g_0 + d_{1,1} g_{-2}
$$

$$
= \frac{14}{4\sqrt{2}} \times \frac{3 + \sqrt{3}}{4\sqrt{2}} + \frac{5 - 3\sqrt{3}}{4\sqrt{2}} \times \frac{1 - \sqrt{3}}{4\sqrt{2}}
$$

$$
+ \frac{2\sqrt{3}}{4\sqrt{2}} \times \frac{-3 + \sqrt{3}}{4\sqrt{2}} + \frac{1 - \sqrt{3}}{4\sqrt{2}} \times \frac{-1 - \sqrt{3}}{4\sqrt{2}} = 2
$$

4.3.3. Finite Discrete Signal Handling with Wavelet Transform

The fast wavelet transforms (4.23) and (4.24) as well as the reconstruction algorithm (4.27) are defined for infinite signal sequences and the length of the filters $H = \{h_k | k = \ldots, -2, -1, 0, 1, 2, \ldots\}$ and $G = \{g_k | k = \ldots, -2, -1, 0, 1, 2, \ldots\}$ may also be infinite, although most of the known wavelet filters have an finite support. However, any signals obtained from equipment is of finite length. Hence there exists a problem when dealing with a finite discrete signal.

For the sake of simplicity or in practice, we will assume that the support of filters $H = \{h_k | k = \ldots, -2, -1, 0, 1, 2, \ldots\}$ and $G = \{g_k | k = \ldots, -2, -1, 0, 1, 2, \ldots\}$ is finite, denoted by L. This is reasonable. Actually, most wavelet functions have filters with finite lengths. The filters of many other wavelet functions have rapidly decreasing properties; thus, when $|k|$ is sufficiently large, one has $h_k \approx 0$ and $g_k \approx 0$. Moreover, without any confusion the indices of filter components are supposed to be from 0 to $L - 1$; thus other components h_k and g_k are zeros for $k \neq 0, 1, \ldots, L - 1$. As a consequence of this assumption, one can write

$$H = \{h_k\}_{k=0}^{L-1} = \{h_0, h_1, \ldots, h_{L-1}\} \quad \text{and}$$
$$G = \{g_k\}_{k=0}^{L-1} = \{g_0, g_1, \ldots, g_{L-1}\}$$

Generally the observed signal $\mathbf{c}_0$ is assumed to be of finite length. Without loss of generality, such a signal with finite length N is denoted by

$$\mathbf{c}_0 = \{c_{0,k}\}_{k=0}^{N-1} = \{c_{0,0}, c_{0,1}, \ldots, c_{0,N-1}\}$$

Let us consider the decomposition algorithm (4.23), with an analysis similar to that in Equation (4.24). In Section 4.3.1 we interpreted this algorithm as convolution between the signal $\mathbf{c}_0$ and the filter $H^* = \{h_{-k} | k = \ldots, -2, -1, 0, 1, 2, \ldots\} = \{\ldots, h_2, h_1, h_0, h_{-1}, h_{-2}, \ldots\}$ in the case of level $j = 1$. In other words, each component in signal $\mathbf{c}_1$ is the inner product (the sum of products of components by components) of signal $\mathbf{c}_0$ and filter series $H = \{h_k | k = \ldots, -2, -1, 0, 1, 2, \ldots\}$ but with two-step shifting (right-hand forward) of components of H once. For example, we have

$$c_{1,0} = \sum_{k=0}^{L-1} h_k c_{0,k} \tag{4.28}$$

and

$$c_{1,1} = \sum_{k=0}^{L-1} h_k c_{0,k+2} \tag{4.29}$$

and so on. The component $c_{1,0}$ in (4.28) can be interpreted as average of vector $\mathbf{c}_0 = (c_{0,0}, c_{0,1}, \ldots, c_{0,N-1})$ of length N with weights $H = \{h_0, h_1, \ldots, h_{L-1}\}$ (assume that $N > L$), and the component $c_{1,1}$ in (4.29) is the average of vector $\mathbf{c}_0 = (c_{0,0}, c_{0,1}, \ldots, c_{0,N-1})$ but generated by moving window of weights H two steps righthand forward (for a discussion of moving-window filtering, see Chapter 2). Other components in $\mathbf{c}_1$ can be computed similarly. However, there are some difficulties while computing $c_{1,-1}$ by

$$c_{1,-1} = \sum_{k=0}^{L-1} h_k c_{0,k-2} \qquad (4.30)$$

which is needed to reconstruct $c_{0,0}$ by the reconstruction algorithm (4.27). In fact, we should at least have $c_{1,-L/2}, \ldots, c_{1,-2}$ and $c_{1,-1}$ to reconstruct $c_{0,0}$, assuming that L is even. The difficulty in implementing (4.30) is that we have no information about $c_{0,-2}$ and $c_{0,-1}$. In order to obtain $c_{1,-L/2}$, information about $c_{0,-L}, \ldots, c_{0,-2}$ and $c_{0,-1}$ should be provided. The similar situation occurs at the other end of $\mathbf{c}_0$.

The basic method to overcome the difficulty is to extend the signal $\mathbf{c}_0$ with a length L (in fact, $L - 2$ is sufficient) at both sides. Let us still denote the extended signal by $\mathbf{c}_0$

$$\mathbf{c}_0 = \{\tilde{c}_{0,-L}, \ldots, \tilde{c}_{0,-1}, c_{0,0}, c_{0,1}, \ldots, c_{0,N-1}, \tilde{c}_{0,N}, \ldots, \tilde{c}_{0,N-1+L}\}$$

where the $\tilde{c}_{0,k}$ values are components to be added. It is interesting to notice that whatever the extension $\tilde{c}_{0,k}$ terms are, the decomposition performed using scheme (4.27) is perfect and the original signal can be recovered using inverse fast wavelet transform (4.27) from

$$\mathbf{c}_1 = \{c_{1,-L/2}, \ldots, c_{1,-1}, c_{1,0}, c_{1,1}, \ldots, c_{1,N/2-1}\}$$

and

$$\mathbf{d}_1 = \{d_{1,-L/2}, \ldots, d_{1,-1}, d_{1,0}, d_{1,1}, \ldots, d_{1,N/2-1}\}$$

where N is assumed to be even. This assumption is fixed in the following discussion.

At the second level of fast wavelet transform, signal $\mathbf{c}_1 = \{c_{1,-L/2}, \ldots, c_{1,-1}, c_{1,0}, c_{1,1}, \ldots, c_{1,N/2-1}\}$ should be similarly extended in order to obtain perfect reconstruction at level 2 and so on. The length of $\mathbf{c}_j$ and $\mathbf{d}_j$ depends on the length of filter, the length of original signal $\mathbf{c}_0$, and the number of levels.

In actual applications, there exist many sophisticated methods for side extension. Typical methods for extending a signal include zero padding, symmetrization, extrapolation, and periodic extension.

Zero Padding. This method assumes that the signal is zero outside the original support. For example, if the original finite signal is

$$\mathbf{c}_0 = \{1, -2, 3, 2, 1, 3\}$$

with length 6, then one of possible zero-padded signals is

$$\mathbf{c}_0^{\text{zero}} = \{0, 0, 0, 0, \mathbf{1}, -\mathbf{2}, \mathbf{3}, \mathbf{2}, \mathbf{1}, \mathbf{3}, 0, 0, 0, 0\}.$$

However, the obvious disadvantage of zero padding is that discontinuities are artificially created at the sides.

Symmetrization. This method assumes that signals can be recovered outside their original support by symmetric boundary-value replication. For example, the signal

$$\mathbf{c}_0 = \{1, -2, 3, 2, 1, 3\}$$

may be extended as

$$\mathbf{c}_0^{\text{sym}} = \{2, 3, -2, 1, \mathbf{1}, -\mathbf{2}, \mathbf{3}, \mathbf{2}, \mathbf{1}, \mathbf{3}, 3, 1, 2, 3\}$$

This is the default mode of the wavelet transform in wavelet toolbox 2 of MATLAB. Symmetrization has the disadvantage of artificially creating discontinuities of the first derivative at the border, but this method works well in general for images.

Extrapolation. To create some values beyond the finite signal $\mathbf{c}_0$, one can employ extrapolation methods. One such method is polynomial extrapolation. Let us consider linear extrapolation. For example, in order to obtain $c_{0,-1}$, first define a line that intersects points $(0, c_{0,0})$ and $(1, c_{0,1})$ and then compute $c_{0,-1}$ such that the point $(-1, c_{0,-1})$ is on the line. Hence we have

$$c_{0,-1} = 2c_{0,0} - c_{0,1}$$

In other words, $c_{0,0}$ is the mean value of $c_{0,-1}$ and $c_{0,1}$. In the similar calculation $c_{0,-2}$, $c_{0,-3}$ as well as $c_{0,N}$, the value of $c_{0,N+1}$ can be determined step by step. Hence in this procedure signal

$$\mathbf{c}_0 = \{1, -2, 3, 2, 1, 3\}$$

may be extended as

$$\mathbf{c}_0^{\text{line}} = \{13, 10, 7, 4, \mathbf{1}, -\mathbf{2}, \mathbf{3}, \mathbf{2}, \mathbf{1}, \mathbf{3}, 5, 7, 9, 11\}$$

Periodization. This method assumes that signal has a period of its length N. The signal may be extended by repeating its series of components. For example, under periodization, signal

$$\mathbf{c}_0 = \{1, -2, 3, 2, 1, 3\}$$

may be extended as

$$\mathbf{c}_0^{per} = \{3, 2, 1, 3, \mathbf{1}, \mathbf{-2}, \mathbf{3}, \mathbf{2}, \mathbf{1}, \mathbf{3}, 1, -2, 3, 2\}$$

in which the first extended components [3,2,1,3] are repeated by the last four components of the original signal and the last extended components [1,−2,3,2] are repeated from the first four components of the original signal. If there is a large difference between the first and last components of the original signal, it is obvious that discontinuity at extended points will be introduced. One method to deal with this problem is to implement translation–rotation transformation (TRT)

$$c_{0,k}^{TRT} = c_{0,k} - c_{0,0} - \frac{k}{N-1}(c_{0,N-1} - c_{0,0}) \qquad (4.31)$$

where k ($=0, 1, \ldots, N-1$) is the index and N is the length of the signal. After the TRT treatment, both $c_{0,0}^{TRT}$ and $c_{0,N-1}^{TRT}$ have the same value 0. Then $\mathbf{c}^{TRT}$ is extended periodically. By direct computation, the signal

$$\mathbf{c}_0 = \{1, -2, 3, 2, 1, 3\}$$

is converted into

$$\mathbf{c}_0^{TRT} = \{0, -3.4, 1.2, -0.2, -1.6, 0\}$$

by TRT through Equation (4.31), then periodically extended as

$$\mathbf{c}_0^{per} = \{1.2, -0.2, -1.6, 0, \mathbf{0}, \mathbf{-3.4}, \mathbf{1.2}, \mathbf{-0.2}, \mathbf{-1.6}, \mathbf{0}, 0, -3.4, 1.2, -0.2\}$$

Conclusion. In the following discussion we assume that the periodic extension is chosen. Under periodic extension we can prove that both approximation signal $\mathbf{c}_1$ and detail signal $\mathbf{d}_1$ given by (4.23) and (4.24) respectively are also periodic with a period of $N/2$. It means that we only have to store $N/2$ components for $\mathbf{c}_1$ and $\mathbf{d}_1$, respectively. Denote by $\mathbf{c}_1 = \{c_{1,0}, c_{1,1}, \ldots, c_{1,N/2-1}\}$ and $\mathbf{d}_1 = \{d_{1,0}, d_{1,1}, \ldots, d_{1,N/2-1}\}$ at the first-level wavelet transform. Generally, at any level j, the approximation $\mathbf{c}_j$

and detail $\mathbf{d}_j$ from the fast wavelet transform are defined as

$$\mathbf{c}_j = \{c_{j,0}, c_{j,1}, \ldots, c_{j,N/2^j-1}\}$$

and

$$\mathbf{d}_j = \{d_{j,0}, d_{j,1}, \ldots, d_{j,N/2^j-1}\}$$

Under periodic extension both decomposition [Eqs. (4.23) and (4.24)] and reconstruction [Eq. (4.27)] are very simple. Let us take $j = 1$ as an example and that assume N and L are even numbers:

1. First, extend signal $\mathbf{c}_0 = \{c_{0,k}\}_{k=0}^{N-1}$ into

$$\mathbf{c}_0^{\text{ext}} = \{c_{0,0}, c_{0,1}, \ldots, c_{0,N-1}, \ldots, c_{0,0}, c_{0,1}, \ldots, c_{0,L-3}\}$$

   with length $(N + L - 2)$.

2. Implement a moving filter (see Chapter 2) along with $\mathbf{c}_0^{\text{ext}}$ by filter $H = \{h_0, h_1, \ldots, h_{L-1}\}$ but move two steps at each time, up to $N/2$ times. The result of such moving filter is taken as approximation signal $\mathbf{c}_1 = \{c_{1,0}, c_{1,1}, \ldots, c_{1,N/2-1}\}$.

3. Implement another moving filter (see Chapter 2) along with $\mathbf{c}_0^{\text{ext}}$ by filter $G = \{g_0, g_1, \ldots, g_{L-1}\}$ but move two steps at each time, up to $N/2$ times. The result of such moving filter is taken as detail signal $\mathbf{d}_1 = \{d_{1,0}, d_{1,1}, \ldots, d_{1,N/2-1}\}$.

Example 4.7. Consider a signal $\mathbf{c}_0 = \{1, 2, 3, 4\}$. The length of this signal is $N = 4$. Compute the approximation and detail signals given by Daubechies wavelet of second order.

Daubechies wavelet filters of second order are $H = \left\{\frac{1-\sqrt{3}}{4\sqrt{2}}, \frac{3-\sqrt{3}}{4\sqrt{2}}, \frac{3+\sqrt{3}}{4\sqrt{2}}, \frac{1+\sqrt{3}}{4\sqrt{2}}\right\}$ and $G = \left\{-\frac{1+\sqrt{3}}{4\sqrt{2}}, \frac{3+\sqrt{3}}{4\sqrt{2}}, -\frac{3-\sqrt{3}}{4\sqrt{2}}, \frac{1-\sqrt{3}}{4\sqrt{2}}\right\}$ with length $L = 4$.

Thus the extended signal should be of length $N + L - 2 = 6$:

$$\mathbf{c}_0^{\text{ext}} = \{1, 2, 3, 4, 1, 2\}$$

By the moving-filter algorithm the first component of approximation signal is given by [see Eq. (4.28)]

$$c_{1,0} = h_0 c_{0,0} + h_1 c_{0,1} + h_2 c_{0,2} + h_3 c_{0,3} = \frac{5 + \sqrt{3}}{\sqrt{2}}$$

and then by moving the filter two steps, the second component of approximation signal is [see Eq. (4.29)]

$$c_{1,1} = h_0 c_{0,2} + h_1 c_{0,3} + h_2 c_{0,4} + h_3 c_{0,5} = \frac{5 - \sqrt{3}}{\sqrt{2}}$$

Similarly, the moving-filter algorithm with filter G is employed to $\mathbf{c}_0^{ext}$. It turns out that

$$d_{1,0} = g_0 c_{0,0} + g_1 c_{0,1} + g_2 c_{0,2} + g_3 c_{0,3} = 0$$

$$d_{1,1} = g_0 c_{0,2} + g_1 c_{0,3} + g_2 c_{0,4} + g_3 c_{0,5} = \frac{2}{\sqrt{2}}$$

At level 2, the approximation $\mathbf{c}_1 = \left\{ \frac{5+\sqrt{3}}{\sqrt{2}}, \frac{5-\sqrt{3}}{\sqrt{2}} \right\}$ is extended to $\mathbf{c}_1^{ext} = \left\{ \frac{5+\sqrt{3}}{\sqrt{2}}, \frac{5-\sqrt{3}}{\sqrt{2}}, \frac{5+\sqrt{3}}{\sqrt{2}}, \frac{5-\sqrt{3}}{\sqrt{2}} \right\}$ with period 2. The moving filtering (only once) along $\mathbf{c}_1^{ext}$ with H and G are

$$c_{2,0} = h_0 c_{1,0} + h_1 c_{1,1} + h_2 c_{1,2} + h_3 c_{1,3} = 5$$

and

$$d_{2,0} = g_0 c_{1,0} + g_1 c_{1,1} + g_2 c_{1,2} + g_3 c_{1,3} = -\sqrt{3}.$$

This gives the final approximation and detail signal with period 1.

Now we will discuss the reconstruction algorithm for a signal with finite length. Still assuming that periodic extension has been employed in the decomposition procedure, we can describe the whole algorithm as follows:

1. Given approximation $\mathbf{c}_1 = \{c_{1,0}, c_{1,1}, \ldots, c_{1,N/2-1}\}$ and detail $\mathbf{d}_1 = \{d_{1,0}, d_{1,1}, \ldots, d_{1,N/2-1}\}$ which are generated by fast wavelet transform with filters $H = \{h_0, h_1, \ldots, h_{L-2}, h_{L-1}\}$ and $G = \{g_0, g_1, \ldots, g_{L_2}, g_{L-1}\}$, we obtain the first inverse filters as $H^* = \{h_{L-1}, h_{L-2}, \ldots, h_1, h_0\}$ and $G^* = \{g_{L-1}, g_{L-2}, \ldots, g_1, g_0\}$. Let L be even and denote $L = 2m$.

2. Periodically extend both $\mathbf{c}_1$ and $\mathbf{d}_1$ from the front to $\mathbf{c}_1^{ext}$ and $\mathbf{d}_1^{ext}$ with length of $\frac{N}{2} + \frac{L}{2} - 1$. Insert 0s between the components of both $\mathbf{c}_1^{ext}$ and $\mathbf{d}_1^{ext}$ as well as before the first components and after the last components such that the resulting signals have length of $N + L - 1$. The resultant signals are

$$\tilde{\mathbf{c}}_1 = \{0, c_{1,N/2-L/2-1}, 0, \ldots, 0, c_{1,N/2-1}, 0, c_{1,0}, 0, \ldots, 0, c_{1,N/2-1}, 0\}$$

and

$$\tilde{\mathbf{d}}_1 = \{0, d_{1,N/2-L/2-1}, 0, \ldots, 0, d_{1,N/2-1}, 0, d_{1,0}, 0, \ldots, 0, d_{1,N/2-1}, 0\}$$

3. Implement moving filters (see Chapter 2) along with $\tilde{\mathbf{c}}_1$ and $\tilde{\mathbf{d}}_1$ with H^* and G^*, respectively. Add two results of moving filters together to give the reconstruction signal.

Example 4.8. From approximation signal

$$c_1 = \left\{ \frac{5 + \sqrt{3}}{\sqrt{2}}, \frac{5 - \sqrt{3}}{\sqrt{2}} \right\}$$

and detail signal

$$d_1 = \left\{ 0, \frac{2}{\sqrt{2}} \right\}$$

at level 1, use inverse fast wavelet transform with Daubechies wavelet of the second order to reconstruct the original signal.

Note that $N/2 = 2$ and $L = 4$, the extended and inserted approximation and detail signals, are

$$\tilde{c}^1 = \left\{ 0, \frac{5 - \sqrt{3}}{\sqrt{2}}, 0, \frac{5 + \sqrt{3}}{\sqrt{2}}, 0, \frac{5 - \sqrt{3}}{\sqrt{2}}, 0 \right\}$$

$$\tilde{d}^1 = \left\{ 0, \frac{2}{\sqrt{2}}, 0, 0, 0, \frac{2}{\sqrt{2}}, 0 \right\}$$

The inverse of Daubechies filters are $H^* = \left\{ \frac{1+\sqrt{3}}{4\sqrt{2}}, \frac{3+\sqrt{3}}{4\sqrt{2}}, \frac{3-\sqrt{3}}{4\sqrt{2}}, \frac{1-\sqrt{3}}{4\sqrt{2}} \right\}$ and $G^* = \left\{ \frac{1-\sqrt{3}}{4\sqrt{2}}, -\frac{3-\sqrt{3}}{4\sqrt{2}}, \frac{3+\sqrt{3}}{4\sqrt{2}}, -\frac{1+\sqrt{3}}{4\sqrt{2}} \right\}$. By the algorithm, the sum of the first terms of two moving filtering is

$$c_{0,0} = 0 \times \frac{1 + \sqrt{3}}{4\sqrt{2}} + \frac{5 - \sqrt{3}}{\sqrt{2}} \frac{3 + \sqrt{3}}{4\sqrt{2}} + 0 \times \frac{3 - \sqrt{3}}{4\sqrt{2}} + \frac{5 + \sqrt{3}}{\sqrt{2}} \frac{1 - \sqrt{3}}{4\sqrt{2}}$$

$$+ 0 \times \frac{1 - \sqrt{3}}{4\sqrt{2}} + \frac{2}{\sqrt{2}} \frac{-3 + \sqrt{3}}{4\sqrt{2}} + 0 \times \frac{3 + \sqrt{3}}{4\sqrt{2}} + 0 \times \frac{-1 - \sqrt{3}}{4\sqrt{2}}$$

$$= 1$$

Similarly, we can compute $c_{0,1}$, $c_{0,2}$, and $c_{0,3}$, and so on.

On the basis of the periodic assumption for a signal, the length of approximation signal $c_{j+1} = \{c_{j+1,0}, c_{j+1,1}, \ldots, c_{j+1,N/2^{j+1}-1}\}$ and detail signal $d_{j+1} = \{d_{j+1,0}, d_{j+1,1}, \ldots, d_{j+1,N/2^{j+1}-1}\}$ is half the of length of c_j. Ordinarily the length of the signal is assumed to be power 2 in the standard fast wavelet transform. For a signal of any length the following "keep one" scheme can be applied:

1. Assume that the length of a finite discrete signal $c_0 = \{c_{0,0}, c_{0,1}, \ldots, c_{0,N-1}\}$ is N. If N is even, then implement the fast wavelet transform once to get approximation signal c_1 and detail signal d_1 with

length $N/2$. If N is odd, then keep the last component $c_{0,N-1}$ and implement fast wavelet transform on $\{c_{0,0}, c_{0,1}, \ldots, c_{0,N-2}\}$ to get the approximation signal $\mathbf{c}_1$ and detail signal $\mathbf{d}_1$;

2. At each level j, denote the length of the approximation signal $\mathbf{c}_j = \{c_{j,0}, c_{j,1}, \ldots, c_{j,N_j-1}\}$ by N_j. If N_j is even, then implement the fast wavelet transform on $\mathbf{c}_j$ to get the approximation signal $\mathbf{c}_{j+1}$ and the detail signal $\mathbf{d}_{j+1}$. If N_j is odd, then keep the last component c_{j,N_j-1} and implement the fast wavelet transform on $\{c_{j,0}, c_{j,1}, \ldots, c_{j,N_j-2}\}$ to obtain the approximation signal $\mathbf{c}_{j+1}$ and the detail signal $\mathbf{d}_{j+1}$.

3. Repeat step 2 until to a level J such that the length of $\mathbf{c}_J$ is 1.

4.3.4. Packet Wavelet Transform

In the fast wavelet transform algorithm, the approximation signal $\mathbf{c}_{j-1}$ is split into two parts: a coarser approximation $\mathbf{c}_j$ and a detail $\mathbf{d}_j$ that disappears in coarser approximation at each level j. The information lost between two successive approximations is captured in the detail signal $\mathbf{d}_j$. Then the next step involves splitting the new approximation signal at coarser resolution and the successive details are never reused.

Instead of splitting only the approximation signal, each detail signal (you can consider this signal as a new signal to be transformed by WT) might also be decomposed into two parts using the same approach as in approximation signal splitting in the corresponding wavelet packet situation. The number of signals generated by this recursive procedure at level j increases as double the number of signals at level $j - 1$. You can see that there are only two signals; one for approximation signal and the other one for detail signal at level 1. As both approximation and detail signals are split by fast wavelet transform, there are four signals generated at level 2, and so on. This recursive splitting of signals is depicted in Figure 4.8, which shows

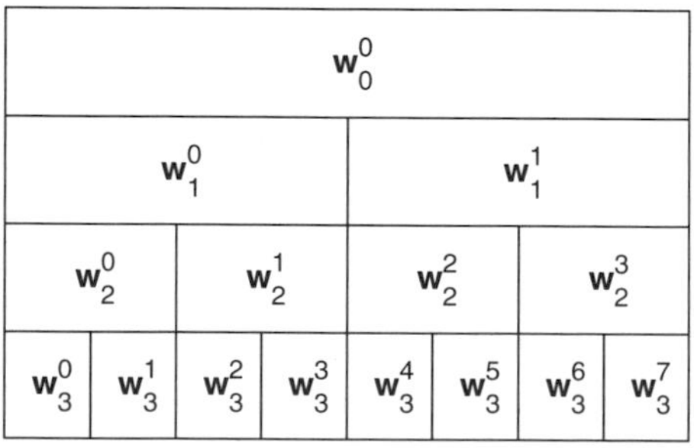

Figure 4.8. The structure of wavelet packet up to three levels.

the structure of a three-levels wavelet packet decomposition. All parts in each line of this diagram are labeled by the level j, and the number p of parts from left to right and are represented as a decomposed signal $\mathbf{w}_j^p = \{w_{j,k}^p | k = \ldots, -2, -1, 0, 1, 2, \ldots\}$. For example, $\mathbf{w}_0^0$ corresponds to an original signal $\mathbf{c}$ at level 0. The idea of this decomposition is to start from a scale-oriented decomposition and then to analyze the obtained signals on frequency subbands.

The computation scheme for wavelet packets is easy when using an orthogonal wavelet. Assume that $\phi(x)$ is a scaling function of a given MRSD and $\psi(x)$ is the corresponding wavelet function associated with a scaling filter $H = \{h_k | k = \ldots, -2, -1, 0, 1, 2, \ldots\}$, and a wavelet filter $G = \{g_k | k = \ldots, -2, -1, 0, 1, 2, \ldots\}$. In the wavelet packet diagram, each decomposed signal $\mathbf{w}_j^p$ at a part can be calculated from the signal corresponding to its abovementioned part as

$$w_{j,k}^{2p} = \sum_{m=-\infty}^{+\infty} h_{m-2k} \, w_{j-1,m}^p$$

$$w_{j,k}^{2p+1} = \sum_{m=-\infty}^{+\infty} g_{m-2k} \, w_{j-1,m}^p$$

Thus, $\mathbf{w}_j^{2p}$ and $\mathbf{w}_j^{2p+1}$ are, respectively, the approximation signal and detail signal of $\mathbf{w}_{j-1}^p$ from one-step fast wavelet transform. In Figure 4.8, each part, except for the parts in the bottom line, covers two subparts that are obtained from themselves by fast wavelet transform in one-step. In general, suppose that an original signal $\mathbf{w}_0^0$ is of finite length, say, N; then each part $\mathbf{w}_j^p$ $(p = 0, 1, \ldots, 2^j - 1)$ should be of finite length $\frac{N}{2^j}$.

At the reconstruction stage each part can be reconstructed from its subparts by inverse fast wavelet transform as

$$w_{j,k}^p = \sum_{m=-\infty}^{+\infty} w_{j+1,m}^{2p} h_{k-2m} + \sum_{m=-\infty}^{+\infty} w_{j+1,m}^{2p+1} g_{k-2m}. \qquad (4.32)$$

It is obvious that the overall diagram of $\mathbf{w}_j^p$ is redundant for reconstruction of the original signal $\mathbf{c}_0 = \mathbf{w}_0^0$. One question to be answered is how many resulting signals $\mathbf{w}_j^p$ should be stored in order to reconstruct perfectly the original signal $\mathbf{w}_0^0$ at the top of the diagram without any redundant information. In order to answer this question, first let us note that all the information contained in any parts of wavelet packet is conserved in its own two subparts because of the orthogonal fast wavelet transform. Hence, if any part in a wavelet packet diagram is chosen to be stored, then any subparts as well as all sub-subparts and its upper parts do not have to be kept. Figure 4.9

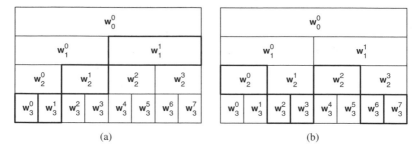

Figure 4.9. The parts to be kept.

illustrates two examples of this kind. The parts that should be kept are given in the thick boxes. In Figure 4.9a the parts kept are the coefficients of standard wavelet transform, i.e., $\{\mathbf{w}_3^0 = \mathbf{c}_3, \mathbf{w}_3^1 = \mathbf{d}_3, \mathbf{w}_2^1 = \mathbf{d}_2, \mathbf{w}_1^1 = \mathbf{d}_1\}$. In Figure 4.9b, the parts kept consist of $\{\mathbf{w}_2^0, \mathbf{w}_3^2, \mathbf{w}_3^3, \mathbf{w}_2^2, \mathbf{w}_3^7, \mathbf{w}_3^6\}$. The reconstruction process can be implemented from these parts as follows. From the bottom parts $\mathbf{w}_3^2$ and $\mathbf{w}_3^3$, the detail signal $\mathbf{w}_2^1$ can be reconstructed by (4.32) with $\mathbf{w}_2^3$ from the parts $\mathbf{w}_3^6$ and $\mathbf{w}_3^7$. Furthermore, $\mathbf{w}_1^0$ and $\mathbf{w}_1^1$ can be reconstructed from $\mathbf{w}_2^0, \mathbf{w}_2^1, \mathbf{w}_2^2$, and $\mathbf{w}_2^3$, and finally $\mathbf{w}_0^0$ from $\mathbf{w}_1^0$ and $\mathbf{w}_1^1$.

4.4. BIORTHOGONAL WAVELET TRANSFORM

The requirement that the translates of scaling and wavelet functions constitute orthonormal bases for the subspaces forming a MRSD and a wavelet decomposition of signal space can be quite restrictive. In theory it has been proved that no wavelet function occurs simultaneously with properties such as smoothness, compact support, symmetry, and translation orthogonality. The Haar wavelet is an example of a wavelet that is symmetric and compactly supported but discontinuous. Sometimes it might be desirable to preserve the other properties and give up orthogonality.

4.4.1. Multiresolution Signal Decomposition of Biorthogonal Wavelet

In order to introduce MRSD for biorthogoal wavelet, let us modify the concept of MRSD by replacing the last requirement [see Eq. (4.15)] with $\{\phi(x - k)|k = \ldots, -2, -1, 0, 1, 2, \ldots\}$ as a Riesz basis for S_0. From now on we say that a scaling function $\phi(x)$ forms a MRSD in the sense of the version modified above.

Mathematically, MRSD for biorthogonal wavelets starts from two multiresolution signal decompositions. It is hard for a student with a chemistry orientation to understand this. Please bear in mind the following fact: If an coordinate system is not orthogonal, then the projection of a point in the system (coordinate value) can be calculated by the inner product of the point vector with the axis vectors of a coordinate system biorthogonal to the original coordinate system. Take a close look at Equation (4.21). At this time, $\psi(x - k)$ are not orthogonal to each other; thus the wavelet coefficients d_{jk} cannot be computed by $\langle f(x), \psi_{j,k}(x)\rangle$ but by $\langle f(x), \tilde{\psi}_{j,k}(x)\rangle$ with a biorthogonal system $\tilde{\psi}_{j,k}(x)$ to $\psi_{j,k}(x)$. However, at this stage, we don't know both $\psi(x)$ and $\tilde{\psi}(x)$.

In order to find both $\psi(x)$ and $\tilde{\psi}(x)$, we take two MRSDs $\{S_j | j = \ldots, -2, -1, 0, 1, 2, \ldots\}$ and $\{\tilde{S}_j | j = \ldots, -2, -1, 0, 1, 2, \ldots\}$ generated by two scaling functions $\phi(x)$ and $\tilde{\phi}(x)$, respectively. Two scaling functions satisfy the biorthogonality conditions

$$\langle \phi(x - k_1), \tilde{\phi}(x - k_2)\rangle = \delta_{k_1 k_2}$$

The scaling functions, both $\phi(x)$ and $\tilde{\phi}(x)$, satisfy their own scaling equations (4.16) with scaling filters $\{h_k | k = \ldots, -2, -1, 0, 1, 2, \ldots\}$ and $\{\tilde{h}_k | k = \ldots, -2, -1, 0, 1, 2, \ldots\}$, respectively. By a long mathematical story, it can be proved that, with the definition

$$g_k = (-1)^{1-k}\tilde{h}_{1-k} \quad \text{and} \quad \tilde{g}_k = (-1)^{1-k}h_{1-k}$$
$$\text{for any} \quad k = \ldots, -2, -1, 0, 1, 2, \ldots \tag{4.33}$$

two wavelet functions $\psi(x)$ and $\tilde{\psi}(x)$ can be created, respectively, by Equation (4.19); and $\psi_{j,k}(x)$ and $\tilde{\psi}_{j',k'}(x)$ are biorthogonal to each other. Thus any signal $f(x) \in L^2(\mathbb{R})$ has two possible decompositions in these bases:

$$f(x) = \sum_{j=-\infty}^{+\infty} \sum_{k=-\infty}^{+\infty} \langle f(x), \psi_{j,k}(x)\rangle \tilde{\psi}_{j,k}(x) \tag{4.34}$$

$$= \sum_{j'=-\infty}^{+\infty} \sum_{k'=-\infty}^{+\infty} \langle f(x), \tilde{\psi}_{j',k'}(x)\rangle \psi_{j',k'}(x) \tag{4.35}$$

Equations (4.34) and (4.35) provide two different decompositions for a signal $f(x)$. In the decomposition (4.34) the decomposition coefficient $c_{j,k} = \langle f(x), \psi_{j,k}(x)\rangle$ is associated with wavelet $\psi_{j,k}(x)$ while the basis is associated with $\{\tilde{\psi}_{j,k}(x)\}$. In fact, the following fast biorthogonal wavelet transform and reconstruction show that the decomposition is implemented by using filters $H = \{h_k | k = \ldots, -2, -1, 0, 1, 2, \ldots\}$ and $G = \{g_k | k = \ldots, -2,$

$-1, 0, 1, 2, \ldots$ } as done in standard fast wavelet transform while the reconstruction is generated with filters $\widetilde{H} = \{ \tilde{h}_k | k = \ldots , -2, -1, 0, 1, 2, \ldots \}$ and $\widetilde{G} = \{ \tilde{g}_k | k = \ldots , -2, -1, 0, 1, 2, \ldots \}$:

Fast Biorthogonal Wavelet Transform

$$c_{j,k} = \sum_{m=-\infty}^{+\infty} h_m \langle f(x), \phi_{j-1, 2k+m}(x) \rangle = \sum_{m=-\infty}^{+\infty} h_{m-2k} c_{j-1, m}$$

$$d_{j,k} = \sum_{m=-\infty}^{+\infty} g_m \langle f(x), \phi_{j-1, 2k+m}(x) \rangle = \sum_{m=-\infty}^{+\infty} g_{m-2k} c_{j-1, m}$$

Fast Inverse Biorthogonal Wavelet Transform

$$c_{j,k} = \sum_{m=-\infty}^{+\infty} c_{j+1, m} \tilde{h}_{k-2m} + \sum_{m=-\infty}^{+\infty} d_{j+1, m} \tilde{g}_{k-2m}$$

Also from Equation (4.35) we can see that a signal $f(x)$ can be decomposed with filters $\{ \tilde{h}_k | k = \ldots , -2, -1, 0, 1, 2, \ldots \}$ and $\{ \tilde{g}_k | k = \ldots , -2, -1, 0, 1, 2, \ldots \}$, and then reconstructed with filters $\{ h_k | k = \ldots , -2, -1, 0, 1, 2, \ldots \}$ and $\{ g_k | k = \ldots , -2, -1, 0, 1, 2, \ldots \}$.

4.4.2. Biorthogonal Spline Wavelets

In this section we introduce biorthogonal wavelets with a minimum size support for a specified number of vanishing moments. These biorthogonal wavelets with symmetric or antisymmetric supports are constructed by Cohen and co-workers [8] based on spline MRSD. These biorthogonal wavelets are known as CDF (Cohen–Daubechies–Feauveau) biorthogonal wavelets or spline biorthogonal wavelets.

In the construction of orthogonal spline wavelets, the scaling function $\phi(x)$ is defined through the orthogonalized version of Fourier transform of the B-spline function. Now this time let us directly choose the B-spline function of degree $m - 1$ as a scaling function $\phi_m(x)$ whose Fourier transform is

$$\hat{\phi}(\omega) = e^{-i(\nu \omega/2)} \left(\frac{\sin \frac{\omega}{2}}{\frac{\omega}{2}} \right)^m$$

where $\nu = 0$ for m even and $\nu = 1$ for m odd.

Cohen, Daubechies, and Feauveau gave the formulas for the biorthogonal filter $\widetilde{H}$ with minimal length.

Table 4.4 gives the filter coefficients for some m and $\tilde{m}$. The resulting scaling functions and wavelet functions for $m = 3$ and $\tilde{m} = 9$ are shown in Figure 4.10.

Table 4.4. Coefficients of Scaling Filters and Biorthogonal Scaling Filters Corresponding to Biorthogonal Spline Wavelets

$p, \tilde{p}$	k	h_k	$\tilde{h}_k$
$m = 2$	0	0.707106781	0.966747552
$\tilde{m} = 6$	$1, -1$	0.353553391	0.447466010
	$2, -2$		-0.169871356
	$3, -3$		-0.107723299
	$4, -4$		0.046956310
	$5, -5$		0.013810679
	$6, -6$		-0.006905340
$m = 3$	$0, 1$	0.530330086	0.942125701
$\tilde{m} = 9$	$-1, 2$	0.176776695	0.002050023
	$-2, 3$		-0.320191968
	$-3, 4$		0.012300136
	$-4, 5$		0.099134782
	$-5, 6$		-0.014112788
	$-6, 7$		-0.020618913
	$-7, 8$		0.005060319
	$-8, 9$		0.002039233
	$-9, 10$		-0.000679744

4.4.3. A Computing Example

Fast biorthogonal wavelet transform is the same as standard fast wavelet transform, while inverse fast biorthogonal wavelet transform is implemented with two dual wavelet filters $\widetilde{H}$ and $\widetilde{G}$. Yet, the computation structure is consistent with that of fast wavelet transform. Thus the strategy introduced in Section 4.3.3 for dealing with signals of finite length still be valid for the fast biorthogonal wavelet transform and reconstruction.

In this section, we give an example demonstrating how to implement such algorithms in the case of biorthogonal spline wavelet. Consider biorthogonal spline wavelet of order $(m, \tilde{m}) = (2, 4)$. The lowpass filter H and its dual $\widetilde{H}$ are

$$H = \left\{ h_{-1} = \frac{\sqrt{2}}{4}, h_0 = \frac{\sqrt{2}}{2}, h_1 = \frac{\sqrt{2}}{4} \right\}$$

$$\widetilde{H} = \left\{ \tilde{h}_{\pm 4} = \frac{3\sqrt{2}}{128}, \tilde{h}_{\pm 3} = -\frac{3\sqrt{2}}{64}, \tilde{h}_{\pm 2} = -\frac{\sqrt{2}}{8}, \tilde{h}_{\pm 1} = \frac{19\sqrt{2}}{64}, \tilde{h}_0 = \frac{45\sqrt{2}}{64} \right\}$$

The relations between filters and their duals [Eq. (4.33)] give

$$\tilde{g}_0 = -h_1 = -\frac{\sqrt{2}}{4}, \quad \tilde{g}_1 = h_0 = \frac{\sqrt{2}}{2}, \quad \tilde{g}_0 = -h_{-1} = -\frac{\sqrt{2}}{4}$$

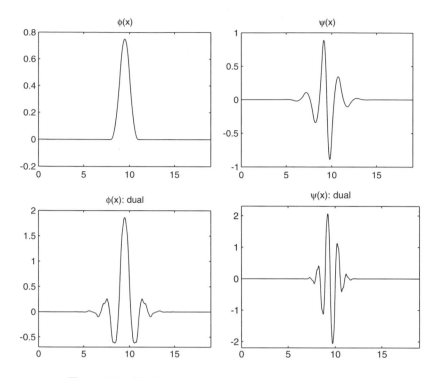

Figure 4.10. Biorthogonal spline scaling functions and wavelets.

and

$$g_{-3} = h_4 = \frac{3\sqrt{2}}{128}, \qquad g_{-2} = -h_3 = \frac{3\sqrt{2}}{64}, \qquad g_{-1} = h_2 = -\frac{\sqrt{2}}{8}$$

$$g_0 = -h_1 = -\frac{19\sqrt{2}}{64}, \qquad g_1 = h_0 = \frac{45\sqrt{2}}{64}, \qquad g_2 = -h_{-1} = -\frac{19\sqrt{2}}{64}$$

$$g_3 = h_{-2} = -\frac{\sqrt{2}}{8}, \qquad g_4 = -h_{-3} = \frac{3\sqrt{2}}{64}, \qquad g_5 = h_{-4} = \frac{3\sqrt{2}}{128}$$

Construct four filters as follows

$$H = \left\{ 0, 0, 0, 0, \frac{\sqrt{2}}{4}, \frac{\sqrt{2}}{2}, \frac{\sqrt{2}}{4}, 0, 0, 0 \right\}$$

$$G = \left\{ \frac{3\sqrt{2}}{128}, \frac{3\sqrt{2}}{64}, -\frac{\sqrt{2}}{8}, -\frac{19\sqrt{2}}{64}, \frac{45\sqrt{2}}{64}, -\frac{19\sqrt{2}}{64}, -\frac{\sqrt{2}}{8}, \frac{3\sqrt{2}}{64}, \frac{3\sqrt{2}}{128}, 0 \right\}$$

$$\tilde{H} = \left\{ \frac{3\sqrt{2}}{128}, -\frac{3\sqrt{2}}{64}, -\frac{\sqrt{2}}{8}, \frac{19\sqrt{2}}{64}, \frac{45\sqrt{2}}{64}, \frac{19\sqrt{2}}{64}, -\frac{\sqrt{2}}{8}, -\frac{3\sqrt{2}}{64}, \frac{3\sqrt{2}}{128}, 0 \right\}$$

$$\tilde{G} = \left\{ 0, 0, 0, 0, -\frac{\sqrt{2}}{4}, \frac{\sqrt{2}}{2}, -\frac{\sqrt{2}}{4}, 0, 0, 0 \right\}$$

where 0s are added such that all the filters have same length $L = 10$.
   Consider a signal of length $N = 10$

$$\mathbf{c}^0 = \{1, 2, 3, 4, 5, 6, 7, 8, 9, 10\}$$

By the periodic extension technique given in Section 4.3.3, it turns out that

$$\mathbf{c}^0_{ext} = \{1, 2, 3, 4, 5, 6, 7, 8, 9, 10, \vdots\ 1, 2, 3, 4, 5, 6, 7, 8\}$$

Employing two-step moving filtering between $\mathbf{c}^0_{ext}$ and H gives

$$c_{1,0} = 5 \times \frac{\sqrt{2}}{4} + 6 \times \frac{\sqrt{2}}{2} + 7 \times \frac{\sqrt{2}}{4} = \frac{24\sqrt{2}}{4}$$

$$c_{1,1} = 7 \times \frac{\sqrt{2}}{4} + 8 \times \frac{\sqrt{2}}{2} + 9 \times \frac{\sqrt{2}}{4} = \frac{32\sqrt{2}}{4}$$

$$c_{1,2} = 9 \times \frac{\sqrt{2}}{4} + 10 \times \frac{\sqrt{2}}{2} + 1 \times \frac{\sqrt{2}}{4} = \frac{30\sqrt{2}}{4}$$

$$c_{1,3} = 1 \times \frac{\sqrt{2}}{4} + 2 \times \frac{\sqrt{2}}{2} + 3 \times \frac{\sqrt{2}}{4} = \frac{8\sqrt{2}}{4}$$

$$c_{1,4} = 3 \times \frac{\sqrt{2}}{4} + 4 \times \frac{\sqrt{2}}{2} + 5 \times \frac{\sqrt{2}}{4} = \frac{16\sqrt{2}}{4}$$

Similarly, two-step moving filtering between $\mathbf{c}^0_{ext}$ and G gives

$$d_{1,0} = 0, \qquad d_{1,1} = -\frac{15\sqrt{2}}{64}, \qquad d_{1,2} = \frac{35\sqrt{2}}{64}$$

$$d_{1,3} = -\frac{225\sqrt{2}}{64}, \qquad d_{1,4} = \frac{45\sqrt{2}}{64}$$

Hence, after fast biorthogonal wavelet wavelet transform at level 1, the approximation and detail signals are, respectively

$$\tilde{\mathbf{c}}_1 = \left\{ \frac{24\sqrt{2}}{4}, \frac{32\sqrt{2}}{4}, \frac{30\sqrt{2}}{4}, \frac{8\sqrt{2}}{4}, \frac{16\sqrt{2}}{4} \right\}$$

$$\tilde{\mathbf{d}}_1 = \left\{ 0, -\frac{15\sqrt{2}}{64}, \frac{35\sqrt{2}}{64}, -\frac{225\sqrt{2}}{64}, \frac{45\sqrt{2}}{64} \right\}$$

with period 5.

In order to reconstruct $\mathbf{c}_0$ from $\mathbf{c}_1$ and $\mathbf{d}_1$, these subsignals are extended as

$$\tilde{\mathbf{c}}_1^{\text{ext}} = \left\{ 0, \frac{32\sqrt{2}}{4}, 0, \frac{30\sqrt{2}}{4}, 0, \frac{8\sqrt{2}}{4}, 0, \frac{16\sqrt{2}}{4}, 0, \frac{24\sqrt{2}}{4}, 0, \right.$$
$$\left. \frac{32\sqrt{2}}{4}, 0, \frac{30\sqrt{2}}{4}, 0, \frac{8\sqrt{2}}{4}, 0, \frac{16\sqrt{2}}{4}, 0 \right\}$$

and

$$\tilde{\mathbf{d}}_1^{\text{ext}} = \left\{ 0, -\frac{15\sqrt{2}}{64}, 0, \frac{35\sqrt{2}}{64}, 0, -\frac{225\sqrt{2}}{64}, 0, \frac{45\sqrt{2}}{64}, 0, 0, 0, -\frac{15\sqrt{2}}{64}, 0, \right.$$
$$\left. \frac{35\sqrt{2}}{64}, 0, -\frac{225\sqrt{2}}{64}, 0, \frac{45\sqrt{2}}{64}, 0 \right\}$$

By the algorithm given in Section 4.3.3, the first component by one-step moving filtering between $\tilde{\mathbf{c}}_1^{\text{ext}}$ and $\tilde{H}$ is $\frac{289}{64}$ and the first component of moving filtering between $\tilde{\mathbf{d}}_1^{\text{ext}}$ and $\tilde{G}$ is $-\frac{225}{64}$. It is obvious that the sum of two components gives the first component of original signal $c_{0,0} = 1$. You can further calculate other components $c_{0,k}$ $(k = 1, \ldots, 9)$ by moving filtering.

4.5. TWO-DIMENSIONAL WAVELET TRANSFORM

4.5.1. Multidimensional Wavelet Analysis

A tensor product wavelet orthonormal basis of $L^2(\mathbb{R}^2)$ is constructed from tensor products of a scaling function $\phi(x)$ and a wavelet function $\psi(x)$. We assume that the scaling function $\phi(x)$ is associated with a one-dimensional MRSD $S = \{S_j | j = \ldots, -2, -1, 0, 1, 2, \ldots\}$ of $L^2(\mathbb{R})$ and the corresponding wavelet function is $\psi(x)$. From this MRSD we construct a two-dimensional (2D) MRSD $S^2 = \{S_j^2 | j = \ldots, -2, -1, 0, 1, 2, \ldots\}$ of $L^2(\mathbb{R}^2)$ by setting $S_j^2 = S_j \otimes S_j$ for each j, where $S_j \otimes S_j$ is the subset of all the functions generated by the products of the functions in S_j. Define a bivariate function by

$$\Phi(x, y) = \phi(x)\phi(y)$$

and its scale and translation versions by

$$\Phi_{j;k,l}(x, y) = \frac{1}{2^j} \Phi\left(\frac{x - 2^j k}{2^j}, \frac{y - 2^j l}{2^j} \right)$$

Then the family $\{\Phi_{j;k,l}(x, y) | k, l = \ldots, -2, -1, 0, 1, 2, \ldots\}$ will be an orthonormal basis for S_j^2.

Denote W_j^2 the detail subspace equivalent to the orthogonal complement of the lower-resolution approximation space S_j^2 in S_{j-1}^2:

$$S_{j-1}^2 = S_j^2 \oplus W_j^2$$

The orthogonal basis of detail subspaces W_j^2 will be called a *two-
dimensional wavelet*.

As in the case of 1D MRSD, it is interesting to construct such a 2D
wavelet basis in W_j^2. For this purpose, let us define three wavelet functions

$$\Psi^1(x, y) = \phi(x)\psi(y), \quad \Psi^2(x, y) = \psi(x)\phi(y),$$
$$\Psi^3(x, y) = \psi(x)\psi(y)$$

(4.36)

and their scale and translation versions for $p = 1,2,3$ and

$$\Psi_{j;k,l}^p(x, y) = \frac{1}{2^j} \Psi^p\left(\frac{x - 2^j k}{2^j}, \frac{y - 2^j l}{2^j}\right)$$

It has been proven, in standard wavelet theory, that the family

$$\{\Psi_{j;k,l}^1(x, y), \Psi_{j;k,l}^2(x, y), \Psi_{j;k,l}^3(x, y) | j, k, l = \ldots, -2, -1, 0, 1, 2, \ldots\}$$

consists of an orthonormal basis for $L^2(\mathbb{R}^2)$.

In other words, the preceding discussion shows that one can easily construct 2D wavelets from any given orthogonal scaling function $\phi(x)$ and the
corresponding wavelet function $\psi(x)$. The three basic wavelets defined in
Equation (4.36) can extract detail at different scales and orientations. As
we know, a scaling function $\phi(x)$ is used mainly to extract low-frequency
information from a signal while the wavelet analyzes high-frequency information. Thus $\psi^1(x, y)$ will extract the information of a two-dimensional
signal, say, an image, in low horizontal (x-axis) frequencies and high
vertical (y-axis) frequencies, whereas $\psi^2(x, y)$ extracts the information
in high horizontal and low vertical frequencies and $\psi^3(x, y)$ extracts the
information in high horizontal and high vertical frequencies.

Figure 4.11 shows the three bivariate wavelets defined by (4.36) with
the cubic spline scaling function and wavelet.

4.5.2. Implementation of Two-Dimensional Wavelet Transform

The fast wavelet transform algorithm derived in Section 4.3.1 is easily
extended in the case of two dimensions. Let $\mathbf{C}_0 = \{c_{0;k,l} | k, l = \ldots,$
$-2, -1, 0, 1, 2, \ldots\}$ be a discrete two-dimensional signal, called an *image*.

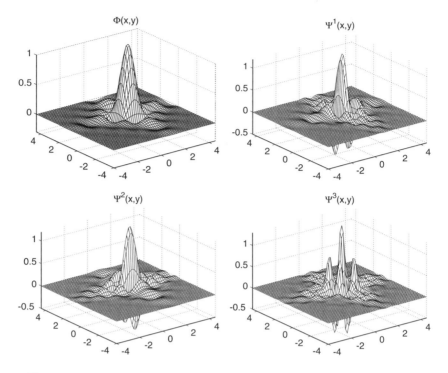

Figure 4.11. Bivariate orthogonal cubic spline scaling function Φ and wavelets Ψ^p.

Associate $\mathbf{C}_0$ with a bivaraite function $f(x, y)$ in space S_0^2 such that $c_{0;k,l} = \langle f(x, y), \Phi_{0;k,l}(x, y) \rangle$. Of course, function $f(x, y)$ can be given by

$$f(x, y) = \sum_{k=-\infty}^{+\infty} \sum_{l=-\infty}^{+\infty} c_{0;k,l} \Phi_{0;,k,l}(x, y)$$

In general, we denote

$$c_{j;k,l} = \langle f(x, y), \Phi_{j;k,l}(x, y) \rangle \quad \text{and} \quad d_{j;k,l}^p = \langle f(x, y), \Psi_{j;k,l}^p(x, y) \rangle$$

Then both $\mathbf{C}_j = \{c_{j;k,l}\}$ and $\mathbf{D}_j^p = \{d_{j;k,l}^p\}$ can be recursively calculated from $\mathbf{C}_0$ by the following fast algorithm:

2D Fast Wavelet Transform

$$c_{j;k,l} = \sum_{m=-\infty}^{+\infty} \sum_{n=-\infty}^{+\infty} h_{m-2k} h_{n-2l} c_{j-1;m,n} \tag{4.37}$$

$$d_{j;k,l}^1 = \sum_{m=-\infty}^{+\infty} \sum_{n=-\infty}^{+\infty} h_{m-2k} g_{n-2l} c_{j-1;m,n} \tag{4.38}$$

$$d^2_{j;k,l} = \sum_{m=-\infty}^{+\infty} \sum_{n=-\infty}^{+\infty} g_{m-2k} h_{n-2l} c_{j-1;m,n} \qquad (4.39)$$

$$d^3_{j;k,l} = \sum_{m=-\infty}^{+\infty} \sum_{n=-\infty}^{+\infty} g_{m-2k} g_{n-2l} c_{j-1;m,n} \qquad (4.40)$$

and, vice versa, $\mathbf{C}_{j-1}$ can be perfectly reconstructed from $\mathbf{C}_j$ and $\mathbf{D}^p_j$, for each j, by the following IFWT.

2D Inverse Fast Wavelet Transform

$$c_{j;k,l} = \sum_{m=-\infty}^{+\infty} \sum_{n=-\infty}^{+\infty} c_{j+1;m,n} h_{k-2m} h_{l-2n} + \sum_{m=-\infty}^{+\infty} \sum_{n=-\infty}^{+\infty} d^1_{j+1;m,n} h_{k-2m} g_{l-2n}$$

$$+ \sum_{m=-\infty}^{+\infty} \sum_{n=-\infty}^{+\infty} d^2_{j+1;m,n} g_{k-2m} h_{l-2n} + \sum_{m=-\infty}^{+\infty} \sum_{n=-\infty}^{+\infty} d^3_{j+1;m,n} g_{k-2m} g_{l-2n}$$

$$(4.41)$$

After J recursive procedures of 2D fast wavelet transform, the original signal $\mathbf{C}_0$ may be decomposed of the following approximation and wavelet signals (also called *subimages*)

$$\{\mathbf{C}_J = \{c_{J;k,l}\}; \mathbf{D}^1_j = \{d^1_{j;k,l}\}; \mathbf{D}^2_j = \{d^2_{j;k,l}\}; \mathbf{D}^3_j = \{d^3_{j;k,l}\} | j = 1, \ldots, J\}$$

We should note that each of $\mathbf{C}_J$, $\mathbf{D}^p_j$ (with $j = J, J-1, \ldots, 1$) is a two-dimensional signal, so we can use an "image" to represent these signals. Thus the resulting structure of 2D fast wavelet transform may be shown by a diagram similar to Figure 4.12, in which a two-level decomposition is given.

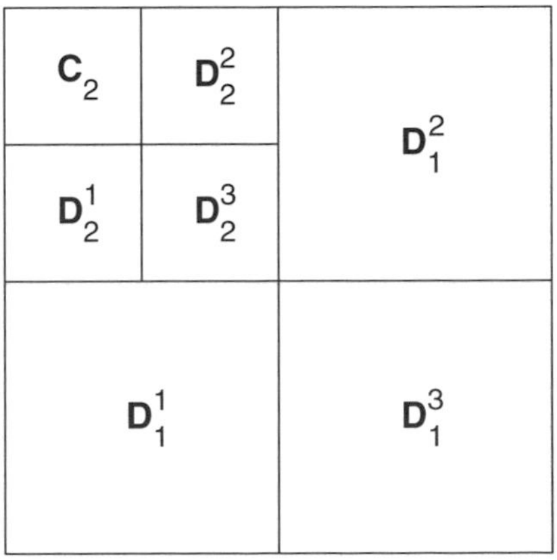

Figure 4.12. The resultant structure after 2D fast wavelet transform at 2 levels.

Although Equations (4.37)–(4.40) appear a little complicated, the 2D fast wavelet transform is in fact very easily implemented. The overall process in each cycle of 2D fast wavelet transform can be divided into two steps. At the level j, the rows of 2D signal $\mathbf{C}_{j-1}$ are firstly transformed by the standard 1D fast wavelet transform algorithm with $H = \{h_k | k = \ldots, -2, -1, 0, 1, 2, \ldots\}$ and $G = \{g_k | k = \ldots, -2, -1, 0, 1, 2, \ldots\}$; and then the columns of the resultant signals are transformed by the same 1D fast wavelet transform, too. The final results are four subsampled-signals $\mathbf{C}_j$, $\mathbf{D}_j^1$, $\mathbf{D}_j^2$, and $\mathbf{D}_j^3$.

In actual applications the 2D signal $\mathbf{C}_0$ is of finite lengths in both row and column directions. Hence we ordinarily consider $\mathbf{C}_0$ as an $M \times N$ matrix where M is the number of rows while N is the number of columns. For such a matrix, 2D fast wavelet transform can be applied to rows and columns, respectively, with the techniques introduced in Section 4.3.3. At the reconstruction stage, 1D inverse fast wavelet transform is applied first to columns and then to rows.

Example 4.9. Apply 2D Haar wavelet transform to the following finite 2D signal:

$$\mathbf{C}_0 = \begin{pmatrix} 1 & 2 & 3 & 4 \\ 4 & 3 & 7 & 8 \\ 6 & 2 & 1 & 8 \\ 2 & 5 & 4 & 7 \end{pmatrix}$$

For the Haar wavelet, we have $h_0 = 1/\sqrt{2}$, $h_1 = 1/\sqrt{2}$, $g_0 = -1/\sqrt{2}$ and $g_1 = 1/\sqrt{2}$. Let us take the first row of $\mathbf{C}_0$ as an example. Using 1D Haar wavelet transform, the approximation coefficients are

$$\frac{1}{\sqrt{2}}(1+2), \qquad \frac{1}{\sqrt{2}}(3+4)$$

and the detail coefficients are

$$\frac{1}{\sqrt{2}}(-1+2), \qquad \frac{1}{\sqrt{2}}(-3+4)$$

The same transform is applied to the other rows of $\mathbf{C}_0$. Arranging the approximation parts of each row transform in the first two columns and the corresponding detail parts in the last two columns results in

$$\begin{pmatrix} 1 & 2 & 3 & 4 \\ 4 & 3 & 7 & 8 \\ 6 & 2 & 1 & 8 \\ 2 & 5 & 4 & 7 \end{pmatrix} \xrightarrow{\text{1D FWT on rows}} \frac{1}{\sqrt{2}} \begin{pmatrix} 3 & 7 & \vdots & 1 & 1 \\ 7 & 15 & \vdots & -1 & 1 \\ 8 & 9 & \vdots & -4 & 7 \\ 7 & 11 & \vdots & 3 & 3 \end{pmatrix}$$

The row approximation information and detail information of the original signal are separated by dotted lines (vertical ellipses) in this equation. The following step is to apply 1D Haar wavelet transform to the columns of the resultant matrix. It turns out that

$$
\frac{1}{\sqrt{2}}
\begin{pmatrix}
3 & 7 & \vdots & 1 & 1 \\
7 & 15 & \vdots & -1 & 1 \\
8 & 9 & \vdots & -4 & 7 \\
7 & 11 & \vdots & 3 & 3
\end{pmatrix}
\xrightarrow{\text{1D FWT on columns}}
\frac{1}{2}
\begin{pmatrix}
10 & 22 & \vdots & 0 & 2 \\
15 & 20 & \vdots & -1 & 10 \\
\cdots & \cdots & & \cdots & \cdots \\
4 & 8 & \vdots & -2 & 0 \\
-1 & 2 & \vdots & 7 & -4
\end{pmatrix}
$$

Thus we have

$$
\mathbf{C}_1 = \begin{pmatrix} 10 & 22 \\ 15 & 20 \end{pmatrix} \qquad
\mathbf{D}_1^1 = \begin{pmatrix} 4 & 8 \\ -1 & 2 \end{pmatrix}
$$

$$
\mathbf{D}_1^2 = \begin{pmatrix} 0 & 2 \\ -1 & 10 \end{pmatrix} \qquad
\mathbf{D}_1^3 = \begin{pmatrix} -2 & 0 \\ 7 & -4 \end{pmatrix}
$$

REFERENCES

1. F. T. Chau, A. K. M. Leung, and J. B. Gao, "A review on application of wavelet transform techniques in chemical analysis: 1989–1997," *Chemomtr. Intell. Lab. Syst.* **43**:165–184 (1998).

2. F. T. Chau, A. K. M. Leung, and J. B. Gao, "Wavelet transform: A novel method for derivative calculation in analytical chemistry," *Anal. Chem.* **70**:5222–5229 (1998).

3. F. T. Chau, A. K. M. Leung, J. B. Gao, and T. M. Shih, "Application of wavelet transform in infared spectrometry: Spectral compression and library search," *Chemometr. Intell. Lab. Syst.* **43**:69–88 (1998).

4. F. T. Chau and A. K. M. Leung, "Application of wavelet transform in electrochemical studies," in *Wavelets in Chemistry*, B. Walczak, ed., Elsevier Science, Amsterdam, 2000, pp. 225–239.

5. F. T. Chau and A. K. M. Leung, "Application of wavelet transform in processing chromatographic data," in *Wavelets in Chemistry*, B. Walczak, ed., Elsevier Science, Amsterdam, 2000, pp. 205–223.

6. F. T. Chau and A. K. M. Leung, "Application of wavelet transform in spectroscopic studies," in *Wavelets in Chemistry*, B. Walczak, ed., Elsevier Science, Amsterdam, 2000, pp. 241–261.

7. C. K. Chui, *An Introduction to Wavelets*, Academic Press, Boston, 1992.

8. A. Cohen, I. Daubechies, and J. C. Feauveau, "Biorthogonal bases of compactly supported wavelets," *Commun. Pure Appl. Math.* **45**:485–560 (1992).

9. I. Daubechies, "Orthonormal bases of compactly supported wavelets," *Commun. Pure Appl. Math.* **41**:909–996 (1988).

10. I. Daubechies, *Ten Lectures on Wavelets*, Vol. 61 of CBMS-NSF Reg. Conf. Ser. Appl. Math., SIAM Press, Philadelphia, 1992.

11. J. B. Gao, F. T. Chau, T. M. Shih, and C. K. Chan, "Application of the fast wavelet transform method to compress ultraviolet-visible spectra," *Appl. Spectrosc.* **50**:339–349 (1996).

12. T. M. Shih, F. T. Chau, J. B. Gao, and J. Wang, "Compression of infrared spectral data using the fast wavelet transform method," *Appl. Spectrosc.* **51**:649–659 (1997).

13. F. T. Chau, J. B. Gao, and T. M. Shih, "Application of the fast wavelet transform method to compress ultraviolet-visible spectra," *Appl. Spectrosc.* **20**:85–90 (1996).

14. S. Mallat, "Multiresolution approximations and wavelet orthogonormal bases of $L^2(\mathbb{R})$," *Trans. Amer, Math. Soc.*, **315**:69–87 (1989).

15. S. Mallat, *A Wavelet Tour of Signal Processing*, Academic Press, San Diego, 1998.

16. Y. Meyer, "Principe d'incertitude, bases hilbertiennes et algèbres d'opérateurs," in *Séminaire Bourbaki*, Vol. 662, Paris, 1986.

CHAPTER

5

APPLICATION OF WAVELET TRANSFORM IN CHEMISTRY

The word "transform" is a mathematical term, but it is frequently used in chemical signal processing. This is because a suitable transform could make difficult calculations become easier and complex signals simpler. Traditionally, Fourier transform (FT) plays a very important role in analytical chemistry. The technique involves a mathematical transformation of signals from one form to another one and is commonly used in analytical instrumentation and computational chemistry for data processing. For example, without the FT technique, it is impossible for chemists to have instruments such as FT-IR, and FT-NMR, as well as some of the signal processing methods mentioned in previous chapters.

As mentioned in Chapter 4, wavelet transform (WT), just as any other mathematical transform, aims at transforming a signal from the original domain to another one in which operations on the signal can be carried out more easily, and the inverse transform reverses the processes. In some respects, the WT resembles the well-known Fourier transform in which the sine and cosine are the analyzing functions. The analyzing function of WT is the wavelet, which is a family of functions derived from a basic function, called the *wavelet basis*, by dilation (or scaling) and translation. Therefore, unlike FT, which is localized in the frequency domain but not in the time domain, WT is well localized in both the time (or position) domain and the frequency (or scale) domain. Furthermore, compared with FT, a large number of basis functions are available with WT. Owing to these differences, one of the main features of WT is that it may decompose a signal into its components directly according to the frequency. With proper identification of the scales with frequency, higher-frequency signals can be separated from lower ones, in the sense that it has zoomin and zoomout capability at any frequency. Since WT can focus on any smaller part of a signal, it can be called a mathematical "microscope." Another feature of WT is that the development of signals into the frequency domain can be

Chemometrics: From Basics To Wavelet Transform, edited by Foo-tim Chau, Yi-zeng Liang, Junbin Gao, and Xue-guang Shao. Chemical Analysis Series, Vol. 164. ISBN 0-471-20242-8. Copyright © 2004 John Wiley & Sons, Inc.

constituted with a flexible choice of waveforms as a basis rather than with just the trigonometric functions like FT.

WT became a popular topic in chemistry and other fields of science starting from late 1980s, after the publication of the important papers by I. Daubechies [1] in 1988 and S. G. Mallat [2,3] in 1989, in which compactly supported orthonormal wavelets and fast calculation algorithms were proposed. Several reference books on WT were published in 1992 and afterwards, such as *Ten Lectures on Wavelets* [4], *An Introduction to Wavelets* [5], *Wavelets: A Tutorial in Theory and Application* [6], *Wavelets: Theory, Algorithms and Applications* [7], and *Wavelets: A Mathematical Tool for Signal Processing* [8]. These books provided general information in wavelet theory, algorithms, and applications. More recently, reference books were also published to introduce applications of WT in various fields of chemistry [9–12].

Since 1989, a large number of papers have been published. In these published works, WT was employed mainly for signal processing in different fields of analytical chemistry that include flow injection analysis (FIA), high-performance liquid chromatography (HPLC), capillary electrophoresis (CE), infrared spectrometry (IR), ultraviolet–visible (UV–vis) spectrometry, mass spectrometry (MS), nuclear magnetic resonance spectrometry (NMR), electroanalytical chemistry, and X-ray diffraction. WT has also been employed to solve certain problems in quantum chemistry and chemical physics. Around 75% of the published works are related to the application of WT in analytical signal processing, which includes data compression, data denoising and smoothing, baseline and background correction, resolution of multicomponent overlapping signals, and analytical image processing. The remaining 25% are related to quantum chemistry, chemical physics, and related fields.

In this chapter, the principles and applications of the WT in data compression, denoising and smoothing, baseline and background removal, resolution enhancement, spectral calibration and regression, classification, and pattern recognition will be explained in detail with examples provided. Then, combined techniques of WT and other chemometric methods will be briefly introduced. The chapter concludes with a review of applications of WT in various chemistry fields.

5.1. DATA COMPRESSION

Nowadays, chemical instruments are usually connected with microcomputer for control of the device, data acquisition, signal processing, interpretation, and reporting of the analyzed results. There is also a growing tendency to combine different chemical devices together to form "hyphenated instruments" for multidimensional measurements. These modern analytical instruments produce multidimensional data that give more information on the analytes. However, this demands larger storage capacities for the instruments, especially when libraries or databases are used for matching, such as infrared (IR) spectroscopy, mass spectroscopy (MS), and nuclear magnetic resonance (NMR) spectroscopy. An alternative to reduce the storage space and the processing time is through signal compression.

In order to store chemical data efficiently, the compression method must not be too computationally demanding. Furthermore, the discrepancy between the original dataset and the data reconstructed from the compressed form should be reasonably small.

Different methods for chemical signal compression have been proposed in the literature and can be classified into two categories. The first one lowers the resolution of the original signal by reducing the number of data to be retained, such as the binary coding method and factor analysis. The second type retains the resolution of the signal through transformation of the signal from one form to another one, such as the well-known FT method. In spite of the availability of these methods, the application of WT on compressing analytical data was proposed and the method was proved to be very efficient.

5.1.1. Principle and Algorithm

In signal processing of chemical datasets, fast wavelet transform (FWT) is usually employed. The algorithm of FWT has been described in the Chapter 4. For a signal c_0, its FWT can be implemented by

$$c_{j,k} = \sum_{m=1}^{N} h_{m-2k} c_{j-1,m} \qquad (5.1)$$

$$d_{j,k} = \sum_{m=1}^{N} g_{m-2k} c_{j-1,m} \qquad (5.2)$$

or simply written as

$$\mathbf{c}_j = H^* \mathbf{c}_{j-1} \qquad (5.3)$$

$$\mathbf{d}_j = G^* \mathbf{c}_{j-1} \qquad (5.4)$$

where c and d with the index represent the elements of the decomposed components $\mathbf{c}$ and $\mathbf{d}$ from the original signal $\mathbf{c}_0$ by WT, $H^* = \{h_{-k}\}_{k \in Z}$ and

$G^* = \{g_{-k}\}_{k \in Z}$ are discrete filters corresponding to the wavelet function $\psi(x)$ and the scaling function $\phi(x)$, and N represents the length of the vector c_{j-1}. In these two equations, the WT of the signal c_0 means that the resulting signals c_1 (called *discrete approximation* or *scale coefficient*) and d_1 (called *discrete detail* or *wavelet coefficient*) are, respectively, the convolution of c_0 with the discrete filters H^* and G^* followed by the property of "downsampling by factor 2."

A schematic diagram showing the operation of the FWT method is shown in Figure 5.1. When FWT is applied to a signal $c_0 = \{c_{0,0}, c_{0,1}, \ldots, c_{0,N-1}\}$ with length $N(= 2^P)$ and P is any positive integer, the scale coefficients c_j and wavelet coefficients d_j at resolution level j are determined by Equations (5.1) and (5.2), respectively. For resolution level $j = 1$, the numbers of the elements of c_1 and d_1 are the same and equal to $N/2$. Then, the decomposition process is applied to c_1 again to obtain the coefficients at the next resolution level. The process is repeated until the desired Jth resolution level is reached. Finally, the original signal is expressed as a collection of the scale and wavelet coefficients in the form of $\{c_J, d_J, d_{J-1}, \ldots, d_1\}$. The total number of coefficients equals the length of the original signal. Because of the downsampling property of the algorithm, it is called the "tree algorithm" or "pyramid algorithm."

The original signal c_0 can be reconstructed from the scale coefficients c_j and wavelet coefficients d_j, $j = J, \ldots, 1$, following the backward procedures of Figure 5.1 using the inverse fast wavelet transform (IFWT)

$$c_{j,k} = \sum_{m=1}^{N} h_{k-2m} c_{j+1,m} + \sum_{m=1}^{N} g_{k-2m} d_{j+1,m} \tag{5.5}$$

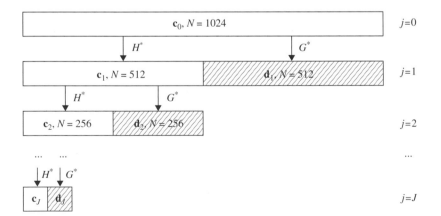

Figure 5.1. A schematic diagram showing the operation of the FWT method on a signal with data length of $N = 1024$.

or simply written as

$$\mathbf{c}_j = H\mathbf{c}_{j+1} + G\mathbf{d}_{j+1} \tag{5.6}$$

where H and G are conjugate filters of the filters H^* and G^*. In the calculation, $\mathbf{c}_j$ and $\mathbf{d}_j$ must followed by a "upsampling by factor 2" with zeros added between each adjacent elements of the vectors.

The decomposition and reconstruction equations presented above are useful for data compression because the WT procedure is capable of retaining a large percentage of the total energy of the signal in the scale coefficients $\mathbf{c}_j$ at different resolution levels. Since the wavelet coefficients are generated from the highpass filter G^*, $\mathbf{d}_j$ reflects the high-frequency information contained in the original data set at the jth level. For most analytical signals, the high-frequency components are usually considered as noise and disposable. Hence, only a small number of the wavelet coefficients is needed to effectively represent the original signal.

For determining which coefficients should be retained, the thresholding method is generally used. Since only the larger coefficients are considered to represent useful information, those coefficients with absolute values greater than a given threshold ε are retained. It is obvious that the choice of the threshold ε affects the compression efficiency and the quality of the reconstructed signal. Usually, a larger ε gives a higher compression ratio but a poorer reconstructed signal. Different procedures for the thresholding operation and the determination of the threshold are discussed in Section 5.1.3.

The general procedure of data compression using WT can be outlined as follows:

1. Apply a WT treatment to the original signal $\mathbf{c}_0$ and to obtain the vector of the scale and wavelet coefficients $\mathbf{w} = \{\mathbf{c}_J, \mathbf{d}_J, \mathbf{d}_{J-1}, \ldots, \mathbf{d}_1\}$ using Equations (5.1) and (5.2).
2. Suppress the small coefficients in $\mathbf{w}$ that are considered too small to contain the useful information of the signal using the thresholding methods. The number of wavelet coefficients to be stored will be decided by the value of the threshold ε.
3. Store the suppressed vector $\mathbf{w}_{\text{store}}$ as the compressed result.
4. Reconstruct the original signal by applying the inverse transform to the $\mathbf{w}_{\text{store}}$ using Equation (5.5) when an original signal is needed.

Example 5.1 illustrates the operation of the above mentioned data compression procedure.

Example 5.1: Compression of a Simulated Signal Using FWT. Figure 5.2, curve (a) shows a simulated signal denoted as a row vector $\mathbf{c}_0$

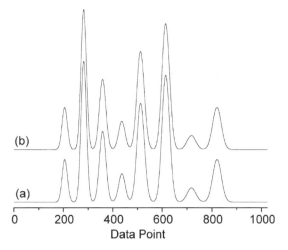

Figure 5.2. Plot of a simulated signal with 1024 data points (a) and the reconstructed signal by WT using 128 coefficients (b).

with 1024 data points. The WT is applied to c_0 with the Daubechies18 ($L = 18$, where L represents the length of the wavelet filters H and G) filters and resolution level $J = 4$, we can obtain the coefficients vector **w** shown in Figure 5.3, plot (a). The vector is in the order of $\{c_{4,(1,\dots,64)}, d_{4,(65,\dots,128)}, d_{3,(129,\dots,256)}, d_{2,(253,\dots,512)}, d_{1,(513,\dots,1024)}\}$. It is clearer in Figure 5.3, plot (b), which is plotted by the absolute values of these coefficients sorted by their magnitudes. It should be noted that WT with different filter and different resolution level J will give a different result; this is discussed

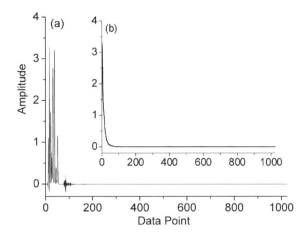

Figure 5.3. Plot of the coefficient vector obtained by applying WT to the simulated signal (a) and its absolute values sorted by magnitude (b).

further in Section 5.1.4. From Figure 5.3, it can be seen that most of the coefficients are small enough to be neglected. The original signal can be represented by only a few WT coefficients.

In order to suppress the small coefficients in **w**, a proper value of the threshold ε is required. In this example, a threshold for retaining 128 coefficients was adopted; that is, only 128 largest coefficients was retained while all other coefficients were set to have value zero. In this way, the compression ratio is 1024/128=8.

Figure 5.2, curve (b) shows the reconstructed signal $\hat{y}_i$ from the 128 retained coefficients as obtained from Example 5.1. Comparing with the original signal y_i in Figure 5.2, curve (a), it can be seen that there is almost no difference between them. The root-mean-square (RMS) error as calculated by $\sqrt{\sum_{i=1}^{N}(\hat{y}_i - y_i)^2/N}$ is only 2.8587×10^{-5}.

Computational Details of Example 5.1

1. Generate the original signal with 1024 data points; refer to curve (a) of Figure 5.2 using the Gaussian equation.
2. Make a wavelet filter—Daubechies18.
3. Set resolution level $J = 4 (10 - 6)$.
4. Perform forward WT to obtain the wavelet coefficients.
5. Perform hard thresholding to the wavelet coefficients keeping the 128 largest coefficients.
6. Construct the signal by applying inverse WT to the 128 retained coefficients.
7. Display Figure 5.2, curves (a) and (b).
8. Display Figure 5.3, plots (a) and (b).
9. Display the RMS error between the original signal and the reconstructed signal.

Note: The MATLAB codes of the examples presented in this chapter are available at the ftp (File Transfer Protocol) server of the publisher (*ftp://www.wiley.com/chemistry*), or by sending an email to the author (*xshao@ustc.edu.cn*).

Most of the programs were developed based on the WaveLab 7.0. The WaveLab is a toolbox developed by the WaveLab Development Team (Jonathan Buckheit, Shaobing Chen, David Donoho, Iain Johnstone, and Jeffrey Scargle) at Stanford University, and it can be downloaded from *http://www-stat.stanford.edu/~wavelab/*. Instructions for installation are also available from the Website. You must install the WaveLab before running these programs.

In the preceding FWT calculation, the length of the original signal must be equal to 2^P. If an odd-number dataset is encountered at a particular resolution level, FWT calculation will be stopped automatically and cannot be processed to the next resolution level. However, in practice, the length of a chemical signal depends on the sampling time (for chromatograms) or wavelength range (for spectra). It is not easy for a chemical instrument to generate 2^P data exactly.

Several methods can be used to cope with the problem, such as the zero padding, symmetrization, extrapolation, and periodization, which have been introduced in Chapter 4. Besides, truncation of data at one or both ends of the original data to the previous power of 2 can also be adopted in some cases. Although these methods are widely used, new ones have also been proposed to improve the FWT algorithm. One of these new methods is called the *coefficient position-retaining* (CPR) method and is described below.

A schematic diagram for applying FWT to a signal with 1231 data with the use of CPR is shown in Figure 5.4. In the CPR approach, if the data length of a scale coefficients c_j at resolution level j, $N_{C,j}$, is an even number, the FWT is applied as usual via Equations (5.1) and (5.2). The scale and wavelet coefficients obtained at resolution $(j+1)$ will have the same length, which is equal to $N_{C,j}/2$. On the other hand, if $N_{C,j}$ is an odd number, FWT is adopted without using the last coefficient of c_j in the calculation. This coefficient is retained and transferred downward to the same position at the next resolution level. It becomes the last coefficient of d_{j+1} at resolution level $(j+1)$. As a result, the scale and wavelet coefficients will have $(N_{C,j} - 1)/2$ and $(N_{C,j} - 1)/2 + 1$ elements, respectively.

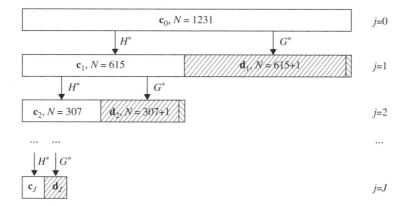

Figure 5.4. A schematic diagram showing the operation of FWT coupled with CPR on a signal with data length of $N = 1231$.

5.1.2. Data Compression Using Wavelet Packet Transform

Wavelet packet transform (WPT) derives from WT. The discrete WT (DWT) is generalized in the WPT procedure to provide a more flexible tool for analytical data analysis. In WT, only partial multiresolution analysis is performed; that is, only c_j is utilized to deduce both the scale and wavelet coefficients at the next resolution level. However, WPT allows a full multiresolution analysis and d_j is also involved at the same time to produce the scale and wavelet coefficients at the next resolution level.

Figure 5.5 shows a schematic diagram for the WPT operation of an analytical signal $\mathbf{w}_0^0$ with a data length of 2^P. In Figure 5.5 and the following text, $\mathbf{w}_j^p$ is used to represent the decomposed component by WPT, where j is the scale parameter and p is an index showing the order of the component in the wavelet packet table. Since WPT is applied to both the scale and wavelet coefficients, the path of WPT is called the *full binary tree* or *WPT binary tree*. In the figure, the original data are decomposed into different components that can be expressed by different bases; that is, the original signal can be expressed by a suitable combination of bases. Generally, a combination of the bases subset is called a *wavelet packet table*. For example, one possible combination of the bases subset to represent the original signal could be $\{\mathbf{w}_3^0, \mathbf{w}_3^1, \mathbf{w}_2^1, \mathbf{w}_2^2, \mathbf{w}_3^6, \mathbf{w}_3^7\}$. Another possible combination could also be $\{\mathbf{w}_2^0, \mathbf{w}_3^2, \mathbf{w}_3^3, \mathbf{w}_2^2, \mathbf{w}_3^6, \mathbf{w}_3^7\}$. How to find the best wavelet packet table, called *best-basis selection*, is discussed in the next section. It can be seen that, whatever the combination will be, the total number of all these coefficients is equal to that of the original data.

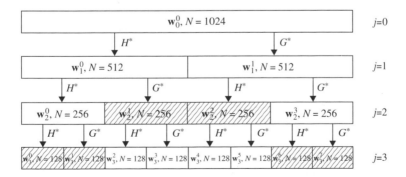

Figure 5.5. A schematic diagram showing the operation of a three-level WPT of a signal with data length of $N = 1024$.

The calculation of the WPT decomposition can be implemented using equations similar to Equations (5.1)–(5.4) with

$$w_{j,k}^{2p} = \sum_{m=1}^{N} h_{m-2k}\, w_{j-1,m}^{p} \tag{5.7}$$

$$w_{j,k}^{2p+1} = \sum_{m=1}^{N} g_{m-2k}\, w_{j-1,m}^{p} \tag{5.8}$$

or simply as

$$\mathbf{w}_{j}^{2p} = H^{*}\mathbf{w}_{j-1}^{p} \tag{5.9}$$

$$\mathbf{w}_{j}^{2p+1} = G^{*}\mathbf{w}_{j-1}^{p} \tag{5.10}$$

where $p = 0, \dots, 2^{j-1} - 1$. Signal reconstruction can also be implemented using equations similar to Equations (5.7) and (5.8) as

$$w_{j,k}^{p} = \sum_{m=1}^{N} h_{k-2m}\, w_{j+1,m}^{2p} + \sum_{m=1}^{N} g_{k-2m}\, w_{j+1,m}^{2p+1} \tag{5.11}$$

or simply written as

$$\mathbf{w}_{j}^{p} = H\mathbf{w}_{j+1}^{2p} + G\mathbf{w}_{j+1}^{2p+1} \tag{5.12}$$

In this manner, data compression using WPT involves the following procedures:

1. Apply WPT to the original signal $\mathbf{w}_{0}^{0}$ up to level J and to obtain all the coefficients $\mathbf{w}_{j}^{p}$ for $j = 1, \dots, J$ and $p = 0, \dots, 2^{j-1} - 1$ using Equations (5.7) and (5.8).
2. Find the best basis to express the original signal.
3. Suppress the small coefficients in the best basis that are considered to be too small to contain useful information of the signal by thresholding methods.
4. Store the suppressed coefficients $\mathbf{w}_{store}$ as the compressed result.
5. Reconstruct the original signal by applying the inverse WPT to the $\mathbf{w}_{store}$ using Equation (5.11) when the original signal is needed.

It can be seen the data compression procedure of WPT requires only one more step of finding the best basis. If the data length of the original data is not 2^{P}, the abovementioned methods, including the CPR method, can also be employed in WPT.

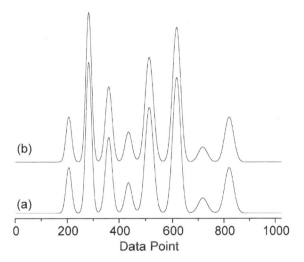

Figure 5.6. Plot of a simulated signal with 1024 data points (a) and the reconstructed by WT signal using 128 coefficients (b).

Example 5.2 gives an application of the procedure listed above for data compression.

Example 5.2: Compression of a Simulated Signal Using WPT. Curve (a) in Figure 5.6 shows the same signal as that in curve (a) of Figure 5.2. The original signal is denoted by the row vector $\mathbf{w}_0^0$. The WPT is applied to the vector with Daubechies18 ($L = 18$) filters and resolution level $J = 9$ to obtain the coefficients vectors $\mathbf{w}_j^p$. It should be also noted that WPT with

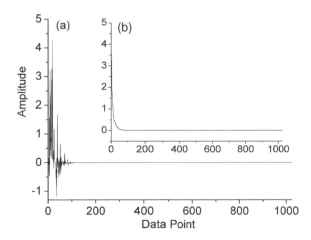

Figure 5.7. Plot of the coefficient vector obtained by applying WT to the simulated signal (a) and its absolute values sorted by magnitude (b).

a different filter and resolution level J will give different result as stated in FWT compression. Compared with FWT again, there is one more step in best-basis selection, and this is discussed in detail in the next section. Plot (a) in Figure 5.7 shows the vector of the coefficients contained in the best basis, which is selected by the Coifman–Wickerhauser entropy method. The remaining procedures are similar to those of the WT compression, that is, to sort the coefficients by their absolute magnitudes [see Fig. 5.7, Plot (b)], to determine a threshold according to the desired compression ratio, to suppress the coefficients whose absolute value is smaller than the threshold, and to store the suppressed coefficients as the compressed result.

In order to compare the performance of the WPT method and FWT compression (Example 5.1), the threshold ε in this example is also assigned for retaining 128 coefficients. The compression ratio is also $1024/128 = 8$. Figure 5.6, curve (b) shows the reconstructed signal from the retained coefficients. The RMS error between the reconstructed signal and the original signal is 4.0842×10^{-5}.

Computational Details of Example 5.2

1. Generate the original signal with 1024 data points; - refer to Figure 5.6, curve (a) by using the Gaussian equation.
2. Make a wavelet filter—Daubechies18.
3. Set resolution level $J = 9$.
4. Perform WPT to obtain the WP coefficients.
5. Find the best basis according to the entropy criteria.
6. Plot the best-basis tree.
7. Apply hard thresholding to the WP coefficients of the best basis keeping the 128 largest coefficients.
8. Construct the signal by applying inverse WPT to the 128 retained coefficients.
9. Display Figure 5.6, curves (a) and (b).
10. Display Figure 5.7, curves (a) and (b).
11. Display the RMS between the original signal and the reconstructed signal.

5.1.3. Best-Basis Selection and Criteria for Coefficient Selection

Considering the principles discussed above, it can be seen that both WT and WPT are useful for data compression because they can turn signals into sparse expansions. Thus, a signal can be accurately represented by only a small number of significant coefficients, and most coefficients have absolute values small enough to be neglected. Furthermore, the overcomplete representation of signal by WPT allows one to choose the appropriate representation (best basis) for the signal compression. However, the following two problems have to be resolved to guarantee the correct use of the FWT or WPT in practical applications:

1. How to find the best basis
2. How to determine the threshold value for discarding "unwanted" coefficients

Methods for Best-Basis Selection. *Best-basis selection* means finding the most efficient basis out of a given set of bases in the wavelet packet tree to represent a given signal. Thus, we intend to find a basis through which some coefficients have high values while the remaining ones have low values; in other words, we wish to differentiate coefficients within a given set as much as possible. One way of selecting an efficient basis from all possible orthonormal bases in the wavelet packet tree is to apply the entropy or information criterion because the amount of information is a measure of inequality of distribution. A basis with coefficients giving more or less the same values would yield a low information or high entropy value. The "best basis" can be defined as the basis giving the minimum entropy or maximum information for its distribution of coefficients. This best-basis definition requires the criterion of the coefficients selection and can also explicitly contain the criterion of the coefficient selection. For instance, the best basis can be specified as the basis with the minimal number of coefficients whose absolute value is higher than the assigned threshold.

In order to calculate the entropy value, several methods have been proposed in the literature. The Coifman–Wickerhauser entropy method and the Shannon–Weaver entropy method are most commonly used.

The Coifman–Wickerhauser entropy is expressed as

$$\lambda_{CW} = -\sum_{m=1}^{M} p_m \log p_m \tag{5.13}$$

where $p_m = (w_{j,m}^p / \|\mathbf{w}_0^0\|)^2$, M is the number of coefficients in the $\mathbf{w}_j^p$, and $\|\mathbf{w}_0^0\|$ is the norm of the original signal, that is, $\sqrt{\sum_{m=1}^{M} |\mathbf{w}_{0,m}^0|^2}$.

The Shannon–Weaver entropy is expressed as

$$H_{SW} = -\sum_{m=1}^{M} q_m \log q_m \tag{5.14}$$

where $q_m = |w_{j,m}^p|^2 / \|\mathbf{w}_j^p\|^2$ and $|w_{j,m}^p|$ is the absolute value of the $w_{j,m}^p$ and $\|\mathbf{w}_j^p\|$ is the RMS norm of the $\mathbf{w}_j^p$. However, Equation (5.14) does not obey the theorem of additive measure of the information. Hence Equation (5.15) is usually used for Shannon–Weaver entropy calculation:

$$\lambda_{SW} = -\sum_{m=1}^{M} |w_{j,m}^p|^2 \log |w_{j,m}^p|^2 \qquad (5.15)$$

Other methods have been proposed for entropy calculation, such as

$$\lambda = \sum_{m=1}^{M} \log p_m \qquad (5.16)$$

where $p_m = |w_{j,m}^p| / \|\mathbf{w}_0^0\|$,

$$\lambda = \sum_{m=1}^{M} (p_m)^k \qquad (5.17)$$

where $p_m = |w_{j,m}^p| / \|\mathbf{w}_0^0\|$, and k is arbitrary parameter,

$$\lambda = \sum_{m=1}^{M} w_{j,m}^p \qquad (5.18)$$

or to calculate the entropy value directly by the number of the coefficients whose absolute value is higher than the predefined threshold. As different methods lead to different results, it is hard to say which one is better. The most suitable method in a case should be chosen by considering the intended application and performing trial calculations.

Once the wavelet packet tree is set up, the entropy of each node (basis) can be determined by any of the methods described above. Then, the best basis can be searched by comparing the entropy values between two adjacent levels. Since each node in the binary tree represents a subspace of the original signal and each subspace is the orthogonal sum of its two children nodes, the search procedure can be performed by a comparison between the entropy value and the sum of entropy values of the two immediate descendants. Nodes with fewer entropy values will be selected as part of the best-basis. For example, in Figure 5.5, if the entropy value of $\mathbf{w}_2^0$ is greater than the sum of the entropy values $\mathbf{w}_3^0$ and $\mathbf{w}_3^1$, the two "children" nodes will be chosen as the best basis. If further decomposition is needed up to level $j = 4$, one should continue to compare the entropy values of $\mathbf{w}_3^0$ and $\mathbf{w}_3^1$ with that of their children nodes $\mathbf{w}_4^0$ and $\mathbf{w}_4^1$, $\mathbf{w}_4^2$ and $\mathbf{w}_4^3$, respectively. However, if the entropy value of $\mathbf{w}_2^1$ is less than the sum of the entropy values $\mathbf{w}_3^2$ and $\mathbf{w}_3^3$, the "parent" node will be chosen as the best basis. Even when decomposition goes on, the comparison stops at this level.

Methods for Determination of Threshold. In both the FWT and WPT compression, it is an important step to suppress the small coefficients by thresholding. A larger ε gives a higher compression ratio but a poorer reconstructed signal. Therefore, determination of the right threshold value is a key step for both compression procedures.

There are many ways to determine the threshold value, and they can be classified as follows:

1. *The universal threshold.* The threshold value of the method is defined as

$$\varepsilon = \sqrt{(2\log_2(S))} \qquad (5.19)$$

where $S = N\log_2(N)$ for the WPT and $S = N$ for the FWT. This threshold must be applied to the data vector **f** that is normalized to the noise level 1. This means that one has to decompose the data to the resolution level $j = 1$ to obtain the detail at the first level, $\mathbf{d}_1$, and estimate the noise based on, for instance, the median of the $\mathbf{d}_1$:

$$s = \frac{median(|\mathbf{d}_1|)}{0.6745} \qquad (5.20)$$

Then, normalize the signal vector **f** by $\mathbf{f}/s$ before performing the data compression.

2. *The threshold defined by the desired compression ratio* (CR). CR is generally defined by N'/N, where N is the data point number of the original signal and N' is the number of the coefficients to be retained. In the cases of a desired CR has been assigned, the threshold can be defined according to the CR value. For example, in Examples 5.1 and 5.2, if the desired CR is assigned as 16, that is, if we must retain 64 largest coefficients, the threshold will have the same value as the 64th coefficient in plot (b) of either Figure 5.3 or 5.7.

3. *The threshold defined by the (root) mean-square error of the recon-structed signal.* This is the most commonly used method because the (root) mean-square error is a measure of the quality of the compression. A large error means that a significant portion of the useful information of the signal is lost after the compression, while an excessively small error will affect the compression efficiency. An effective way to determine the threshold value for the method is by trial and error, which is generally time-consuming. Fortunately, when the basis of the WT is orthonormal, the mean-square error of the reconstructed signal equals the sum of the squared coefficients suppressed from the vector **w**. The quality of the compression can be easily evaluated directly by the suppressed coefficients without reconstruction. In

some cases, the threshold value can also be defined by acceptable local pointwise error of signal reconstruction.

4. *The minimum description length* (MDL) *criterion.* MDL is a criterion for optimization of the retained largest coefficients and the error of the reconstructed signal. Use of the MDL criterion allows an objective threshold selection, which can be particularly useful for real data where the noise level is difficult to estimate. The MDL is defined by

$$MDL(k, n) = \min \left[\frac{3}{2} k \log (N) + \frac{N}{2} \log \|(I - \Theta^k) w_n^k f\|^2 \right] \qquad (5.21)$$

where N denotes the length of the signal f, n describes the filter, $k(0 \le k < N)$ denotes the number of nonzero elements in the vector w_n^k, I is the N-dimensional identity matrix, Θ^k is a thresholding operation that keeps the k largest elements (in absolute value) and sets all other elements to zero, and $\|(I - \Theta^k) w_n^k f\|$ represents the error between the original signal and the reconstructed signal with the k largest elements.

The first term of the MDL cost function can be considered as the penalty function of retaining the nonzero wavelet coefficients, whereas the second term describes the error between the original signal and the reconstructed signal with the k largest wavelet coefficients. These two terms of the MDL represent opposing demands. We would like to have k as small as possible in order to compress the data. At the same time we would like to minimize the distortion between the estimated and the true signal. But a larger k value usually gives a smaller error. By adding the linear penalty function to the log term, we can observe the minimum of the MDL cost function for a small amount of k values. We are not interested in a k near N, because this makes no sense in terms of data compression.

Besides, there are still other methods to determine the threshold value, and some of which are discussed in Section 5.2.2, because they are used for data denoising.

Example 5.3: Data Compression of an Experimental NMR Spectrum.
Figure 5.8 shows a NMR spectrum of a biological sample measured on a Bruker DMX 500 NMR spectrometer. It consists of 32,768 data points. In order to compress the spectrum using WT, we can first represent the spectrum as a vector c_0 and scale its values into the range of 0–1, then perform a FWT to the data vector according to the process described in Section 5.1.1. Figure 5.9 shows the wavelet coefficients obtained with Symmlet10 ($L = 20$) filters and biggest resolution $J = 10$. In order to see clearly, the figure was enlarged and clipped into the magnitude range within ± 0.01. It

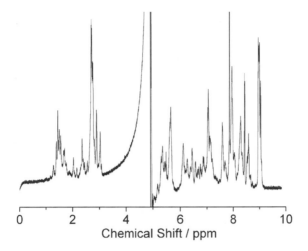

Figure 5.8. An experimental NMR spectrum with 32,768 data points.

can be seen that there are only a few coefficients whose absolute value is greater than 0.001; most of them are small enough to be suppressed.

In order to investigate the effect at different compression ratios, three thresholds for retaining 2048 (CR = 16), 1024 (CR = 32), and 512 (CR = 64) coefficients were set, respectively. After the thresholding operation, the reconstructed signals were as shown in Figure 5.10. It can be seen that there is almost no difference between them. The RMS between the original spectrum and the reconstructed spectra are, respectively, 1.3642×10^{-4}, 1.6977×10^{-4}, and 2.7684×10^{-4}.

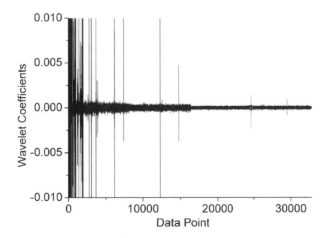

Figure 5.9. An enlarged plot of the wavelet coefficients obtained by applying WT to the experimental NMR spectrum using the filters of Symmlet10 ($L = 20$) and resolution level $J = 9$.

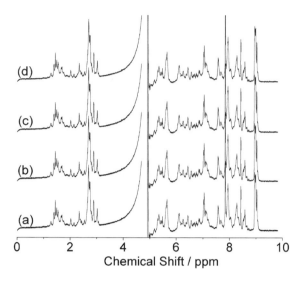

Figure 5.10. Comparison of the experimental NMR spectrum (a) and the reconstructed spectra by WT using 2048 (b), 1024 (c), and 512 (d) coefficients.

Computational Details of Example 5.3

1. Load the experimental NMR signal with 32,768 data points (see Fig. 5.9).
2. Make a wavelet filter—Symmlet10.
3. Set resolution level $J = 10$ $(15 - 5)$.
4. Perform WT to obtain the wavelet coefficients.
5. Apply thresholding to the wavelet coefficients keeping the 2048, 1024, 512 largest coefficients, respectively.
6. Construct the signal by applying inverse WT to the retained coefficients, respectively.
7. Display Figure 5.9.
8. Display Figure 5.10.
9. Display the RMS between the original signal and the reconstructed signals.

Note: Results in Table 5.1 can be obtained by changing the wavelet filters in step 2, and the results in Table 5.2 can be obtained by changing the resolution level in step 3.

As has been stated above, different wavelet filters and different decomposition levels will result in different effects of compression. In order to obtain optimal filter and resolution level J for the spectrum, the RMS

Table 5.1. RMS between Experimental NMR Spectrum and Reconstructed Signal by WT with Different Wavelet Filters and Resolution Level $J = 10$ (CR $= 64$)

Filter	Length (L)	RMS ($\times 10^{-4}$)	Filter	Length (L)	RMS ($\times 10^{-4}$)	Filter	Length (L)	RMS ($\times 10^{-4}$)
Haar	2	7.2549	Daub18	18	2.9404	Sym10	20	2.7684
Daub4	4	3.4853	Daub20	20	2.9512	Coif1	6	3.2101
Daub6	6	2.9214	Sym4	8	2.7955	Coif2	12	2.7339
Daub8	8	2.9050	Sym5	10	2.7627	Coif3	18	2.7040
Daub10	10	2.8948	Sym6	12	2.7224	Coif4	24	2.7332
Daub12	12	2.7940	Sym7	14	2.7498	Coif5	30	2.7773
Daub14	14	2.7958	Sym8	16	2.7171			
Daub16	16	2.9843	Sym9	18	2.7017			

between the original measured spectrum and reconstructed signal by different filters and different resolution levels should be investigated.

Table 5.1 summarizes the RMS by different wavelet filters with decomposition level $J = 10$ and CR $= 64$. Filters including Haar, Daubechies ($L = 4$–20), Symmlet ($L = 8$–20), and coiflet ($L = 6$–30) are investigated. It can be seen that different filters give different RMS results and the Symmlet ($L = 18$) gives a slightly better filter for compression of this NMR spectrum, yet the difference among these filters are not significant.

Table 5.2. RMS between Experimental NMR Spectrum and Reconstructed Signal by WT with Wavelet Filters of Coiflet, Daubechies, and Symmlet and Different Decomposition Level (CR $= 64$)

Filter: Coiflet ($L = 18$)		Filter: Daubechies ($L = 12$)		Filter: Symmlet ($L = 18$)	
Resolution Level J	RMS ($\times 10^{-4}$)	Resolution Level J	RMS ($\times 10^{-4}$)	Resolution Level J	RMS ($\times 10^{-4}$)
6	158.54	6	152.03	6	162.67
7	3.1719	7	3.2537	7	3.1017
8	2.8036	8	2.8865	8	2.7781
9	2.7152	9	2.8089	9	2.7065
10	2.7040	10	2.7940	10	2.7017
11	2.7040	11	2.7940	11	2.7017
12	2.7040	12	2.7940	12	2.7017
13	2.7040	13	2.7940	13	2.7017
14	2.7040	14	2.7940	14	2.7017
15	2.7040	15	2.7940	15	2.7017

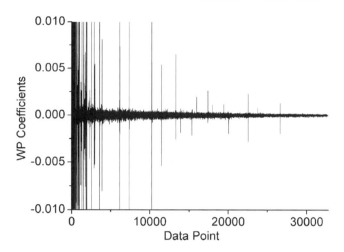

Figure 5.11. An enlarged plot of the WPT coefficients obtained by applying WPT to the experimental NMR spectrum using the filters of Symmlet6 ($L = 12$) and resolution level $J = 10$.

Table 5.2 lists the RMS by coiflet ($L = 18$), Daubechies ($L = 12$), and Symmlet ($L = 18$) filters at different decomposition levels. The CR is also 64. It can be seen that, for all the three filters, the RMS reaches a minimum when decomposition level J is larger than 10. However, the RMS decreases with the increase of the decomposition level within $J = 6$–10.

WPT can also be used for compression of the NMR spectrum. Figure 5.11 shows the coefficients obtained by Symmlet6 filter and decomposition level $J = 10$. The best basis was searched by the Coifman–Wickerhauser entropy method. Figure 5.12 compares the original NMR spectrum and its reconstructed signals from 2048 (CR = 16), 1024 (CR = 32), and 512 (CR = 64) coefficients. The RMS is 1.3649×10^{-4}, 1.7054×10^{-4}, and 2.6616×10^{-4}, respectively.

5.2. DATA DENOISING AND SMOOTHING

Noise in chemical signals is generally defined as the instantaneously irreproducible signals caused by interfering physical or chemical processes, imperfections in the experimental apparatus, and other irregularities, by which the experimental results are often complicated. Therefore, denoising and smoothing is a problem of interest in all fields of science and technology, and a large number of filtering methods have been developed, such as the Fourier filtering method, the Savitsky–Golay smoothing method, and the Kalman filtering method.

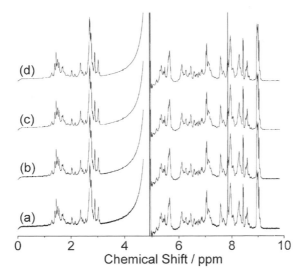

Figure 5.12. Comparison of the experimental NMR spectrum (a) and the reconstructed spectra by WPT using 2048 (b), 1024 (c), and 512 (d) coefficients.

Strictly speaking, denoising removes the small-amplitude components of the transformed signal regardless of the frequency, whereas smoothing minimizes the fluctuation in the signal by removing the high-frequency components regardless of amplitude [see *Anal. Chem.* **69**:78 (1997)].

The underlying philosophy of the denoising by WT is generally thought to resemble the traditional Fourier filtering, in which the high-frequency components are cut off by lowpass filters. But strictly speaking, this kind of filtering should be classified into smoothing according to the definition given above. The principle of WT denoising and smoothing is in fact similar to that of compression by WT, which suppresses the small elements from the coefficient vectors. In the following, both denoising and smoothing are discussed in further detail.

5.2.1. Denoising

According to the principle of the WT denoising, the procedure of denoising with WT can be summarized as follows:

1. Apply a WT to noisy signal f_{noisy} and obtain the wavelet coefficient vector w.
2. Suppress those elements in w that are thought to be attributed to noise by thresholding.
3. Apply the inverse transform to the suppressed w to obtain the denoised signal $f_{denoised}$.

Generally, there are two methods for thresholding operation: hard thresholding and soft thresholding. The policy for hard thresholding is "keep or kill." The absolute values of all transformed coefficients are compared to the fixed threshold value ε. If the magnitude of a coefficient is less than ε, the coefficient is replaced by zero; otherwise, it does not change:

$$w_k^{hard} = \begin{cases} 0 & \text{if } |w_k| < \varepsilon \\ w_k & \text{if } |w_k| \geq \varepsilon \end{cases} \tag{5.22}$$

where w_k represents an element of the coefficients $\mathbf{w}$. Soft thresholding shrinks all the coefficients toward the origin in the following way:

$$w_k^{soft} = \begin{cases} 0 & \text{if } |w_k| < \varepsilon \\ sign(w_k)(|w_k| - \varepsilon) & \text{if } |w_k| \geq \varepsilon \end{cases} \tag{5.23}$$

Thus, if the magnitude of the coefficient is less than ε, the coefficient is replaced by zero; otherwise, a ε is subtracted from the absolute value of the coefficient.

In data compression, only hard thresholding is used because the aim of compression is to remove the small coefficients. But in denoising both thresholding strategies can be used. On the other hand, because the aim of the thresholding differs between compression and denoising, the method determining the threshold value ε for denoising is different from that for compression. Several methods have been proposed, which can be summarized by the following:

1. *Stein's Unbiased Risk Estimate (SURE) Method.* The SURE method is a hard-thresholding approach where the major work is invested in finding the right threshold for different scales. The thresholding is performed on each scale. For calculation of the threshold value ε for scale j, the squared coefficients is first computed by

$$a_k = (w_{j,k})^2 \tag{5.24}$$

where $w_{j,k}$ represents an element of the jth-scale coefficients in $\mathbf{w}$, and sorted in ascending order. Then, the cumulative total of the a_k is computed by

$$b_k = \sum_{i=1}^{k} a_i \tag{5.25}$$

Further, a vector $\mathbf{c}$ with the same size of the coefficient number of the scale j, n_j, is designed by $c_k = n_j - k$; thus, the first element is $n_j - 1$ and the last element is 0. A risk value for every coefficient, r_k, is then computed by

$$r_k = \frac{(n_j - 1) + b_k + a_k c_k}{n_j} \tag{5.26}$$

The coefficient that has the minimum r_k is selected as the threshold value for the scale j:

$$\varepsilon = \min\{r_k\} \tag{5.27}$$

It should be noted that the absolute value of the coefficient is used for the thresholding operation.

2. *Visually Calibrated Adaptive Smoothing Method*. This method is generally called the *VISU* method. The threshold value is computed simply by

$$\varepsilon = (2 \log n)^{1/2} \tag{5.28}$$

where n is the number of the elements in the coefficient vector. This value can be used for hard- or soft-thresholding operation. For methods 1 and 2, the thresholding sometimes is performed only on the coefficients in the index interval $[2^J + 1, n]$ (the wavelet or detail coefficients part) where J is the predefined largest decomposition scale. J must be smaller than $\log_2 (N)$ when the length of the signal is N.

3. *Hybrid Thresholding Method*. This method is called *HYBRID* method because it is a soft-thresholding method where in some cases the VISU threshold $\varepsilon_A = (2 \log n_j)^{1/2}$ is used, and in other cases SURE threshold $\varepsilon_B = \min\{r_k\}$ is used, depending on the parameters e and p defined by

$$\begin{cases} e = \frac{\|\mathbf{w}_j\|^2 - n_j}{n_j} \\ p = \frac{J^{3/2}}{\sqrt{n_j}} \end{cases} \tag{5.29}$$

where J is the largest decomposition scale, $\mathbf{w}_j$ and n_j are the coefficients of the scale j and its length, respectively. If $e < p$, then soft VISU thresholding is performed, if not, the smaller value in ε_A and ε_B is used to perform the soft-thresholding operation. In this method, the thresholding is performed on the coefficients of each scale as in the SURE method.

4. *Median Absolute Deviation (MAD) Method*. The MAD method is a soft-thresholding method. It is also applied to the individual scales. The threshold used for each scale j is the same as in VISU method, i.e., $\varepsilon = (2 \log n)^{1/2}$, but the coefficients must be normalized by $\mathbf{w}_j / s_j$, where $s_j = median(|\mathbf{w}_j|)/0.6745$, before the thresholding is performed.

5. *MINMAX Thresholding Method*. This method finds the optimum threshold for each scale according to the mean-square error between the

original signal and the denoised signal. A set of threshold values is used that satisfy $\varepsilon \leq (2 \log n)^{1/2}$. When the coefficient number n is very large, the optimum threshold ε will approach $(2 \log n)^{1/2}$.

In practice uses, we can use the thresholds provided in the literature. For example, we can use the following threshold suggested by Donoho et al. in the WaveLab (*http://www-stat.stanford.edu/~wavelab/*):

$$\{\varepsilon_j\}_{j=1,\ldots,16} = \{0, 0, 0, 0, 0, 1.27, 1.47, 1.67, 1.86, 2.05, 2.23,$$
$$2.41, 2.6, 2.77, 2.95, 3.13\} \tag{5.30}$$

It should be noted that, for most of these above methods, the threshold must be applied to the data vector normalized to the noise level 1. It would be a good idea to normalize the signal **f** by **f**/s before performing the denoising or smoothing, where s can be computed by Equation (5.21).

Example 5.4: Denoising of a Simulated Chromatogram. Curves (a) in Figures 5.13 and 5.14 are simulated chromatograms with 1024 data point by Gaussian function

$$\mathbf{V} = \sum_{j=1}^{n} c_j \exp\left(-4\ln(2)\left(\frac{t - t_{0,j}}{W_{1/2,j}}\right)^2\right) \tag{5.31}$$

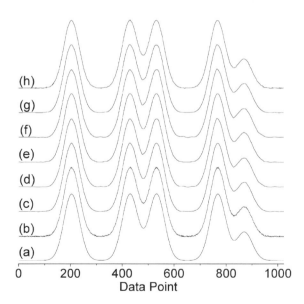

Figure 5.13. Plots of a simulated clean chromatogram (a), noisy chromatogram with random noise of SNR = 100 (b), and the denoised results by WT with hard- (c), soft- (d), SURE (e), VISU (f), HYBRID (g), and MINMAX (h) thresholding methods, respectively.

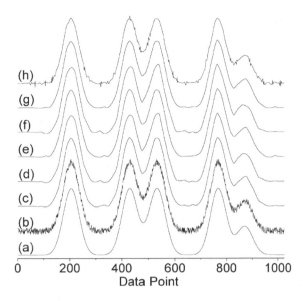

Figure 5.14. Plots of a simulated clean chromatogram (a), noisy chromatogram with random noise of SNR = 20 (b), and the denoised results by hard- (c), soft- (d), SURE (e), VISU (f), HYBRID (g), and MINMAX (h) thresholding methods, respectively.

where **V** is the simulated chromatogram, t is retention time, n is the number of peaks, and c_j, $t_{0,j}$, and $W_{1/2,j}$ are the concentration, position, and full width at half height of the peak j, respectively. Curve (b) in Figure 5.13 shows a noisy chromatogram generated by adding random noise in curve (a). The SNR calculated by the ratio of the maximum of the signal and the maximum of the noise is 100. Curve (b) in Figure 5.14 is generated in the same way, but the SNR is 20.

Curves (c)–(h) in Figures 5.13 and 5.14 are the denoised results achieved by different thresholding methods. The Symmlet4 ($L = 8$) filter and the decomposition level $J = 5$ were adopted in the calculation. Curves (c) and (d) in Figure 5.13 are obtained, respectively, by hard- and soft-thresholding methods with the threshold $\varepsilon = 0.003 \times [\max(\mathbf{w}) - \min(\mathbf{w})]$, which is determined by trial and error. In Figure 5.14, curves (c) and (d) are obtained in the same way as in Figure 5.13, but the threshold $\varepsilon = 0.015 \times [\max(\mathbf{w}) - \min(\mathbf{w})]$ is adopted. The thresholds for all the other thresholding methods are determined by the abovementioned equations. Soft thresholding is adopted for the *HYBRID* method, while hard thresholding operation is adopted for the other methods. In order to compare the denoised results with the original simulated signal, RMS values between curves (a) and (c)–(h) are calculated, respectively. These values are 0.0064, 0.0081, 0.0055, 0.0065, 0.0092, and 0.0060 for the results in

Figure 5.13 (SNR = 100), and 0.0288, 0.0325, 0.0287, 0.0289, 0.0355, and 0.0304 for the results in Figure 5.14 (SNR = 20). It can be seen that all the methods can give acceptable results, but the distortion will increase with the noise level.

Computational Details of Example 5.4

1. Generate the noisy signal with 1024 data points; see curve (b), Figure 5.13 using the Gaussian equation.
2. Make a wavelet filter—Symmlet4.
3. Normalize the data to noise level 1.
4. Set resolution level $J = 5$ $(10 - 5)$.
5. Perform forward WT to obtain the wavelet coefficients.
6. Perform thresholding to the wavelet coefficients to suppress the smaller coefficients, construct the signal by applying inverse WT to the suppressed coefficients, and display the denoised results, respectively, by
   6.1. Hard thresholding with manually determined threshold [Fig. 5.13, (c)].
   6.2. Soft thresholding with manually determined threshold [Fig. 5.13, (d)].
   6.3. SURE (hard) thresholding [Fig. 5.13, (e)].
   6.4. HYBRID (soft) thresholding [Fig. 5.13, (f)].
   6.5. VISU (hard) tresholding [Fig. 5.13, (g)].
   6.6. MINMAX (hard) thresholding [Fig. 5.13, (h)].

Note: The only difference between Figures 5.13 and 5.14 is the SNR of the noisy signal.

From the denoised results curves (c) and (d) in both Figures 5.13 and 5.14, it can be seen that both the hard and soft thresholding by manually determined threshold ε can give satisfactory results. But these results are obtained by trial and error, which is tedious and time-consuming. Among the results of SURE, VISU, HYBRID, and MINMAX, whose thresholds are determined automatically by the abovementioned equations, the SURE method result is the best. It is clear that curves (e) in both Figure 5.13 and 5.14 are clean and the least distorted. Although curves (f) and (g) are also clean enough, there are obvious distortions, especially for the last small peak in the shoulder. From curve (h) in the two figures, it is clear that the MINMAX is not a good method for denoising of the chromatogram, because there is still noise in the denoised result, especially when the SNR is low in Figure 5.14.

5.2.2. Smoothing

According to the principle of the WT smoothing, the procedure of smoothing with WT can be summarized as follows:

1. Apply a WT to noisy signal f_{noisy} and obtain the wavelet coefficients vector **w**.
2. Remove those elements that are thought to represent high-frequency fluctuation in **w**.
3. Apply the inverse transform to the suppressed **w** to obtain the smoothed signal $f_{smoothed}$.

It can be seen that the only difference between denoising and smoothing is the second step. For denoising, coefficients thought to be small enough are suppressed, but for smoothing, the coefficients representing high-frequency information are removed. According to the principles of WT, the detail coefficients at low scale are generally attributed to high-frequency components.

In order to remove the coefficients corresponding to the high-frequency component, a threshold for cutting off those coefficients must be provided for smoothing. Unfortunately, it is not so easy to determine a frequency threshold as in denoising. Therefore, the threshold value is again determined by trial and error.

As indicated in Figure 5.3, 5.7, and 5.9, it is difficult to determine a cutoff position from the figure. Therefore, an improved algorithm was proposed for DWT calculation. It can be described by

$$c_{j,k} = \frac{1}{\sqrt{2}} \sum_{m=1}^{L} h_{j,m} c_{j-1,k-m} \tag{5.32}$$

$$d_{j,k} = \frac{1}{\sqrt{2}} \sum_{m=1}^{L} g_{j,m} c_{j-1,k-m} \tag{5.33}$$

where the h_j and g_j are obtained by inserting $2^{j-1} - 1$ zeros into the every adjacent element of h and g in Equation (5.1) and (5.2). Subsequently, the length of the filter L will be doubled. By this algorithm, both $c_j = \{c_{j,k}\}$

and $\mathbf{d}_j = \{d_{j,k}\}$ keep the same length with $\mathbf{c}_{j-1} = \{c_{j-1,k}\}$. Therefore, we can plot $\mathbf{c}_j$ and $\mathbf{d}_j$, respectively, to inspect their frequency. Furthermore, the algorithm can be used to decompose the signals $\mathbf{c}_0$ with any length without using the special techniques such as coefficient position retaining (CPR) method.

The corresponding reconstruction algorithm can be described by

$$c_{j,k} = \frac{1}{\sqrt{2}} \sum_{m=1}^{L} h_{j,m} c_{j+1,k-m} + \frac{1}{\sqrt{2}} \sum_{m=1}^{L} g_{j,m} d_{j+1,k-m} \qquad (5.34)$$

Example 5.5: Data Smoothing of a Simulated Chromatogram. Figure 5.15 shows the $\mathbf{c}_j$ and $\mathbf{d}_j$ obtained by Equations (5.32) and (5.33) with a Daubechies4 ($L = 4$) filter and $J = 5$ from the simulated signal of curve (b) in Figure 5.14. It is clear that the $\mathbf{d}_1 \sim \mathbf{d}_4$ are attributed to the high-frequency components (noise). Therefore, the smoothed results can be obtained by reconstruction by Equation (5.34) where $\mathbf{d}_1 \sim \mathbf{d}_4$ is set to zeros, which is shown in Figure 5.16. By comparison of the smoothed result with the simulated signal in Figure 5.16, it can be seen that there is almost no distortion after the smoothing. The RMS between the two curves is only 0.0028, which is smaller than any of the denoising methods discussed above.

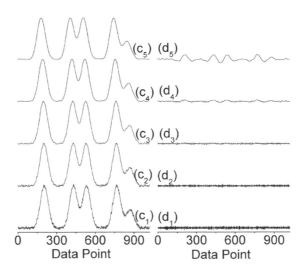

Figure 5.15. Plot of the wavelet coefficients obtained by WT with the improved algorithm.

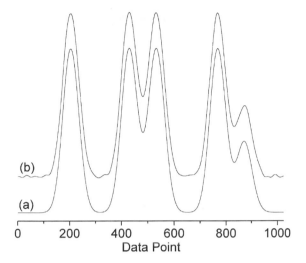

Figure 5.16. Comparison of the simulated clean chromatogram (a) and the smoothed result by WT with the improved algorithm (b).

Computational Details of Example 5.5

1. Generate the noisy signal with 1024 data points [Fig. 5.14 (a),(b)] using the Gaussian equation.
2. Make a wavelet filter—Daubechies4.
3. Set resolution level $J = 5$.
4. Perform WT to obtain the **c** and **d** components with the improved algorithm.
5. Display Figure 5.15.
6. Perform smoothing by replacing the $\mathbf{d}_1 - \mathbf{d}_4$ with zeros and constructing the signal with inverse WT.
7. Display Figure 5.16.
8. Display the RMS between the original signal and the smoothed signal.

We can also propose a method to estimate an approximate scale threshold for deciding the coefficients at which scale should be set to zero using the characteristics of Equations (5.32) and (5.33) and the concept of the Nyquist critical frequency in sampling theory. The Nyquist critical frequency is defined by

$$f_c = \frac{1}{2\Delta_S} \qquad (5.35)$$

where Δ_S is sampling time interval. For a signal with the Nyquist critical frequency f_c, the frequency of the $\mathbf{c}_j$ and $\mathbf{d}_j$ calculated by Equations (5.32) and (5.33) will match the following inequality:

$$\begin{cases} 2^{-j}f_c \le f_{d_j} \le 2^{-j+1}f_c \\ f_{c_j} > 2^{-j}f_c \end{cases} \tag{5.36}$$

Then, if we take the peak of an analytical signal (e.g., a chromatogram) as a periodic signal, following the definition of the Nyquist critical frequency f_c, the frequency of the peak can be defined by

$$f_p = \frac{1}{2W_b} \tag{5.37}$$

where W_b denotes the width of the peak. Because the aim of the smoothing is to remove high-frequency fluctuation and retain the chromatographic signal, the frequency of the coefficients to be removed, f_{cut}, should be much higher than that of the analytical peak:

$$f_p = \frac{1}{2W_b} \ll f_{cut} = \min(f_{d_j}) = 2^{-j}f_c = 2^{-j}\frac{1}{2\Delta s} \tag{5.38}$$

If we define the f_{cut} by

$$f_{cut} = \frac{1}{2W_{th}} \tag{5.39}$$

where W_{th} denotes the estimated maximal width of noise, then rearrange the Equation (5.38) and perform a logarithm on both sides, we obtain

$$\begin{cases} -j + \log_2\left(\frac{1}{2\Delta s}\right) = \log_2\left(\frac{1}{2W_{th}}\right) \\ W_{th} \ll W_b \end{cases} \tag{5.40}$$

Therefore, if we denote the j in Equation (5.40) as j_{th}, we have

$$\begin{cases} j_{th} = \log_2\left(\frac{W_{th}}{\Delta s}\right) \\ W_{th} \ll W_b \end{cases} \tag{5.41}$$

We can estimate an approximate scale threshold j_{th} by Equation (5.41); all the detail coefficients at the scale lower than the j_{th} should be set to zero in the reconstruction calculation. The difficulty in using the method is that the parameter W_{th} must be estimated by experience.

Example 5.6: Data Smoothing of an Experimental Chromatogram. Figure 5.17 shows an experimental chromatogram measured by an reverse-phase HPLC and postcolumn reaction detection with arsenazo III. The sample is composed of six rare-earth ions (Lu, Yb, Tm, Er, Ho, and

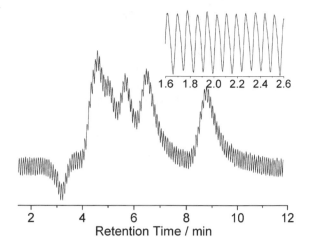

Figure 5.17. Experimental chromatogram with great level noise of a mixed rare-earth solution sample.

Yb). The chromatogram recorded between 1.505 and 11.811 min sampled every 0.005 min is shown. The data length is 2048. The noise is caused by the strong absorption of the postcolumn reaction agent and the pulse of the mobile phase. Therefore, the noise is an oscillation as shown in the enlarged part of the figure. It is impossible to denoise such a chromatogram by other commonly used denoising methods. But we can easily smooth the chromatogram by using the WT smoothing method mentioned above.

At first, we can determine a scale threshold by examination of the enlarged part of the chromatogram. It is easy to find that the maximal width

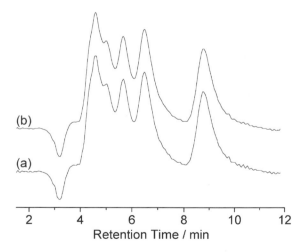

Figure 5.18. Comparison of the smoothed results of the noisy chromatogram by WT with general (a) and the improved (b) algorithms.

of the noise is 16 points (0.08 min). By Equation (5.41), we can calculate the scale threshold $j_{th} = \log_2(0.008/0.005) = \log_2(16) = 4$. Therefore, if we perform a WT decomposition with $J = 4$ and Dabechies4 filter, and then reconstruct the signal with all the detail coefficients ($d_1 \sim d_4$) set to zero, we can obtained the smoothed result, which is shown in Figure 5.18. It should be noted that the scale threshold by Equation (5.41) can be used for both the general algorithm by Equations (5.1), (5.2), and (5.5) and the improved algorithm by Equations (5.32)–(5.34). Curve (a) in Figure 5.18 is obtained by the former algorithm, and curve (b) is obtained by the later one. It can be seen that they are almost the same.

Computational Details of Example 5.6

Mallat algorithm: Figure 5.18 (a):

1. Load the experimental chromatographic signal (Fig. 5.17).
2. Make a wavelet filter—Symmlet4.
3. Set resolution level $J = 5$ $(10 - 5)$.
4. Perform WT to obtain the wavelet coefficients with the Mallat algorithm.
5. Perform smoothing by replacing the detail coefficients of last Four scales with zeros and constructing the signal with inverse WT.
6. Display Figures 5.17 and 5.18 (a).
7. Display the RMS between the original signal and the smoothed signal.

Improved algorithm: Figure 5.18 (b):

1. Load the experimental chromatographic signal (Fig. 5.17).
2. Make a wavelet filter—Daubechies4.
3. Set resolution level $J = 4$.
4. Perform WT to obtain the **c** and **d** components with the improved algorithm.
5. Perform smoothing by replacing the $d_1 - d_4$ with zeros and constructing the signal with inverse WT.
6. Display Figures 5.17 and 5.18 (b).
7. Display the RMS between the original signal and the smoothed signal.

It should be noted that, in Examples 5.4, 5.5, and 5.6, the filters used are not optimized, better results may be obtained if we use better filters.

5.2.3. Denoising and Smoothing Using Wavelet Packet Transform

As discussed in Section 5.1.2, WPT is a development of WT. The only difference between the two methods is the decomposition tree. In WT, only the approximation coefficients at each scale are used for further decomposition, but in WPT, the further decomposition is applied to both the approximation and the detail coefficients. The procedures of data denoising and smoothing using WPT can be summarized as follows:

1. Apply a WPT to the original signal $\mathbf{w}_0^0 = \mathbf{f}_{\text{noise}}$ up to a predefined scale J and to obtain the coefficients $\mathbf{w}_j^p$ for $j = 1, \ldots, J$ and $p = 0, \ldots, 2^{j-1} - 1$ using Equations (5.7) and (5.8).
2. Find the best basis to express the original signal.
3. Suppress the coefficients in best-basis selection that are considered to be attributed to noise by thresholding methods for denoising, or remove those coefficients that are thought to represent high frequency fluctuation for smoothing.
4. Reconstruct the denoised or the smoothed signal by the inverse WPT using Equation (5.11).

As for the thresholding operation in step 2, all the methods used in the WT denoising and smoothing can be employed.

The improved algorithm by Equations (5.32)–(5.34) can also be used for WPT smoothing. Decomposition and reconstruction can be implemented by

$$w_{j,k}^{2p} = \sum_{m=1}^{L} h_{j,m} w_{j-1,k-m}^{p} \tag{5.42}$$

$$w_{j,k}^{2p+1} = \sum_{m=1}^{L} g_{j,m} w_{j-1,k-m}^{p} \tag{5.43}$$

$$w_{j,k}^{p} = \sum_{m=1}^{L} h_{j,m} w_{j+1,k-m}^{2p} + \sum_{m=1}^{L} g_{j,m} w_{j+1,k-m}^{2p+1} \tag{5.44}$$

As in WT, the frequency of the coefficients $\mathbf{w}_j^0$ will be lower than $2^{-j} f_c$, and $\mathbf{w}_j^p$ for $p = 1, \ldots, 2^j - 1$ will be in the range of $p \times 2^{-j} f_c \sim (p+1) \times 2^{-j} f_c$.

Compared with WT smoothing, it is more important for WPT smoothing to develop a scale threshold estimation method because the components of the coefficients are much more complicated. It is impossible to examine all the $\mathbf{w}_j^p$ by trial and error. Unfortunately, it is difficult to derive a simple equation for the determination of both j and p as a threshold. In practical applications, we can simply use Equation (5.41) to estimate a minimum scale j_{th} and use $p_{th} = 0$ as the threshold, that is, to decompose the signal with $J = j_{th}$ to obtain the coefficients $\{\mathbf{w}_j^p\}$ and set all the $\mathbf{w}_j^p$ with $p > 0$ to zero in the reconstruction calculation. It is easy to prove that the result of such an approach will be exactly the same as that of WT smoothing with scale threshold j_{th}. Therefore, if a satisfactory result can be obtained with the threshold j_{th} and $p_{th} = 0$, WPT is not a good choice; it would be better to use WT. We have mentioned that the difference between WPT and WPT is the decomposition tree, and WPT gives a finer decomposition than WT. Therefore, if we cannot obtain a satisfactory result with the threshold j_{th} and $p_{th} = 0$, we should perform the WPT decomposition with $J = j_{th}+2, j_{th}+3$, or an even higher scale, and then estimate an approximate threshold p_{th} by

$$\min(f_{\mathbf{w}_j^p}) = (p+1) \times 2^{-j} f_c = (p+1) \times 2^{-j} \frac{1}{2\Delta_S} = f_{cut} = \frac{1}{2W_{th}} \qquad (5.45)$$

i.e.,

$$p_{th} = 2^j \frac{\Delta_S}{W_{th}} - 1 \qquad (5.46)$$

By such an approach, trials sometimes, may still be needed to determinate the parameter p_{th} in practical applications. It is clear that only those $\mathbf{w}_j^p$ with $j = J$ and p around p_{th}, instead of all the $\mathbf{w}_j^p$, need be examined. (*Note:* The best-basis selection step is not necessary if we use the scale threshold in smoothing.)

Example 5.7: Denoising and Smoothing of Simulated and Experimental Chromatograms Using WPT. Figure 5.19 shows the denoised and the smoothed results of the simulated chromatogram in Figure 5.14, curve (b) using the WPT. The denoised curve (a) in Figure 5.19 is obtained with Symmlet4 filter and $J = 10$ by hard thresholding. The best basis was selected using Coifman–Wickerhauser entropy method, and the threshold value $\varepsilon = 0.008 \times [\max(\mathbf{w}) - \min(\mathbf{w})]$ was determined by trial and error. The smoothed curve (b) in Figure 5.19 is obtained with the same filter and $J = 4$. $j_{th} = 4$ and $p_{th} = 0$ were used as the scale threshold. Both the denoising and the smoothing give us a satisfactory result. If we further compare the denoised and the smoothed results, we can find that the smoothed result is superior to the denoised result.

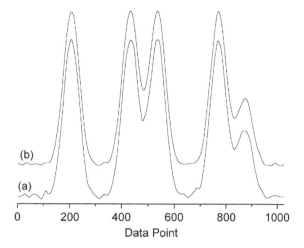

(b)

(a)

0   200   400   600   800   1000

Data Point

Figure 5.19. Comparison of the denoised (a) and smoothed (b) results by WPT of the simulated noisy chromatogram.

Computational Details of Example 5.7

Denoising: Figures 5.19 (a):
1. Generate the original signal with 1024 data points using the Gaussian equation.
2. Make a wavelet filter—Symmlet4.
3. Normalize the data to noise level 1.
4. Set decomposition level $D = 10$ (dyadic length of the signal).
5. Perform WPT to obtain the WP coefficients.
6. Find best basis according to the entropy criteria.
7. Apply hard thresholding to the WP coefficients of the best basis.
8. Perform inverse WPT with the WP coefficients after thresholding to obtain the denoised signal.
9. Display Figure 5.19, curve (a).

Smoothing: Figure 5.19 (b):
1. Generate the original signal with 1024 data points using the Gaussian equation.
2. Make a wavelet filter—Symmlet4.
3. Set scale threshold $J_{th} = 4$ and $P_{th} = 0$, and set decomposition level $J = J_{th}$.
4. Perform WPT to obtain the WP coefficients.
5. Perform inverse WPT with WP coefficients within the scale threshold to obtain the smoothed signal.
6. Display Figure 5.19, curve (b).

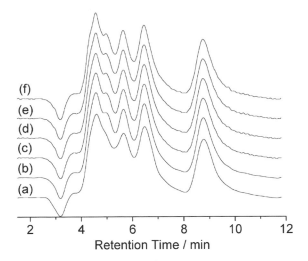

Figure 5.20. Plots of the smoothed chromatograms by WPT with scale threshold of $j_{th} = 7$ and $p_{th} = 2$ (a), 3 (b), 4 (c), 5 (d), 6 (e), and 7 (f).

As mentioned above, if we perform smoothing of the experimental chromatogram in Figure 5.17 using the improved algorithm of WPT by scale threshold $j_{th} = 4$ and $p_{th} = 0$, the smoothed result should be same as the result in Figure 5.18, curve (b). If we want to smooth out the small fluctuation remained in that curve, we can perform a further smoothing using the WPT by scale threshold $j_{th} = 7$ and $p_{th} < 7$, because WPT further decomposes the $\mathbf{w}_4^0$ into $\mathbf{w}_7^0 \sim \mathbf{w}_7^7$ and the f_{cut} for $j_{th} = 4$ and $p_{th} = 0, 1 \times 2^{-4} f_c$, is equal to that for $j_{th} = 7$ and $p_{th} = 7, 8 \times 2^{-7} f_c$.

Figure 5.20 shows the smoothed results by scale threshold $j_{th} = 7$ and $p_{th} = 2 \sim 7$. It is evident that the threshold $j_{th} = 7$ and $p_{th} = 7$ gives the same result with that in Figure 5.18 (b). Going from curve (f) to curve (a), it is clear that the curves become increasingly smooth with the decrease of p_{th}. Therefore, we can conveniently choose a curve as our smoothed result.

5.2.4. Comparison between Wavelet Transform and Conventional Methods

In order to compare the WT or WPT denoising and smoothing with the conventional methods, the simulated and the experimental chromatograms are smoothed by moving-average, Savitsky–Golay, and FFT filtering methods, respectively. Figures 5.21 and 5.22 show their results. The smoothing window or filter width for the three methods is respectively 25, 13, and 13 points for the simulated chromatogram in Figures 5.21 and 25, 17, and 17 points for the experimental chromatogram in Figure 5.22. By comparing these

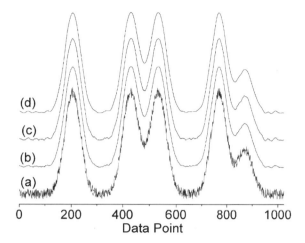

Figure 5.21. A simulated noisy chromatogram (SNR = 20) (a) and the smoothed results with the use of moving-average (b), Savitsky–Golay (c), and FFT (d) filtering.

curves with those by WT or WPT, it can be found that, for the simulated signal, all the methods can give similar results, but for the experimental signal, WT and WPT give more satisfactory results.

5.3. BASELINE/BACKGROUND REMOVAL

In many cases, the baseline drift or background in an analytical signal is just like the noise, which often increases the difficulties in further processing. The baseline drift mainly induces errors in the determination

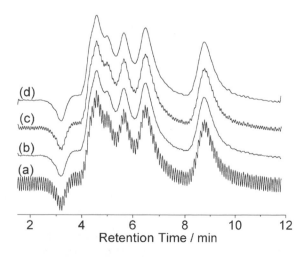

Figure 5.22. The experimental chromatogram (a) and the smoothed results with the use of moving-average (b), Savitsky–Golay (c), and FFT (d) filtering.

of the peak height and peak area, which are very important parameters for signals analysis. The background always blurs the analytical signals. It is difficult or even impossible to analyze a signal with a strong background. In practice, an artificial baseline or background is usually drawn beneath the peak, although the error cannot be completely eliminated by this method. Therefore, most chemists prefer to find the exact shape of the baseline or background and then subtract it from the original signal. But it is not easy to obtain the "true" baseline or background because they are generally represented by curves instead of linear functions.

5.3.1. Principle and Algorithm

Baseline drift or background interference can be classified as a long-term noise. This property differs from that of common noise in that the frequency of the drifting baseline or background is always quite lower than the signals to be analyzed. In wavelet decomposition, the baseline or background component in an analytical signal should be easy to separate from the drifting signals. The removal procedure is similar to WT denoising and smoothing. The only difference is that the coefficients representing the lower-frequency components are suppressed. The following two methods are generally used.

Method A
1. Decompose the experiment data into discrete approximations c_j and discrete details d_j by Equations (5.1) and (5.2).
2. Examine the c_j by visual inspection to find a c_j that resembles the drifting baseline or background, and denote the scale j as j_{max}. The j_{max} can also be determined by examination of the d_j, because there should be no information of the signal in $d_{j_{max}+1}$.
3. Reconstruct the signal by Equation (5.5) from only those d_j with $j \leq j_{max}$, that is, set $c_{j_{max}}$ to zeros.

Method B
1. Decompose the experiment data into discrete approximations c_j and discrete details d_j by Equations (5.32) and (5.33).
2. Examine the c_j by visual inspection to find a c_j (denote the scale j as j_{max}) that resembles the drifting baseline or background.
3. Subtract the selected $c_{j_{max}}$ from the original signal by $c_0 - c_{j_{max}}$ or sometimes $c_0 - f \times c_{j_{max}}$, where f is an arbitrary factor.

5.3.2. Background Removal

Background removal is a universal problem in spectral studies, such as absorption spectroscopy and reflection spectroscopy. Example 5.8 describes a procedure to remove the background absorption from an EXAFS spectrum using the methods A and B, respectively.

Example 5.8: Background Removal of an EXAFS Spectrum. *Extended X-ray absorption fine structure* (EXAFS) is an X-ray absorption spectrum. A typical experimental absorption spectrum of copper with synchrotron radiation light source is shown in Figure 5.23a. For analyzing the EXAFS spectrum, we must separate the useful information (oscillation part) from the total raw spectrum. Then convert the oscillation part into k space by using the equation $k = [0.263(E - E_0)]^{1/2}$, where E_0 is the absorption edge (the E_0 for Cu is 8393.5 eV). At last, the oscillation signal is filtered by FFT and then fitted by the theoretical equation

$$
\chi(k) = \sum_j \frac{N_j}{kr_j^2} |f_j(k)| e^{-k^2\sigma_j^2} e^{-2r_j/\lambda}
$$
$$
\times \sin\left[2kr_j + \varphi(k) + \frac{0.2625r_j\Delta E_0}{k} \right]
$$

(5.47)

where $\chi(k)$ is the filtered oscillation signal multiplied by a factor k^3, j is the number of the coordination shells, $f_j(k)$ is the amplitude value that can be obtained from handbook, $\varphi(k)$ is the phase displacement of scattering, ΔE_0 is the difference between the theoretical value and the experimental value of the E_0, N is coordination number, r is coordination distance, σ is Debye–Waller factor, and λ is electron mean-free path. The aim is to obtain the structural parameter N, r, σ, and λ. Generally, the cubic spline interpolation method is used for the background removal and the least-squares method is used for the curve fitting. In our experience, it will take several hours to analyze one spectrum.

We can use method A for the background removal of an EXAFS spectrum. Figure 5.24a shows an experimental spectrum of a Cu sample, and panel (b) shows the spectrum in k space. In order to decompose the spectrum into its approximation and details, we can perform a WT on the k-space spectrum with Equations (5.1) and (5.2). Figure 5.25 shows the decomposed results, $\{c_4, d_4, d_3, d_2, d_1\}$, obtained with Daubechies8 ($L = 8$) filter and $J = 4$. It is clear that the information on the EXAFS oscillation is decomposed into the d_j and that c_4 represents the smooth background absorption. Therefore, if we reconstruct the spectrum from the d_j components only, we will obtain the spectrum without the background absorption. The dotted line in Figure 5.24c shows the reconstructed

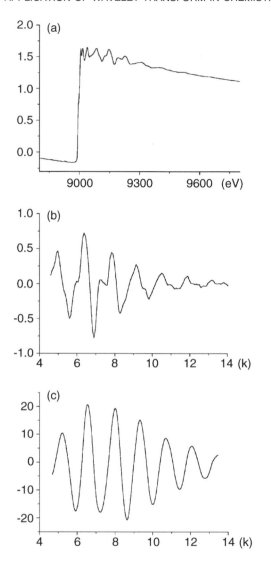

Figure 5.23. General procedures for analying an EXAFS spectrum: (a) raw spectrum; (b) oscillation part; (c) Filtered oscillation signal.

spectrum. From the result, it is obvious that the background is removed. The solid line in Figure 5.24c shows the result obtained by the conventional cubic spline curve fitting after many trials. By comparing the two curves, it can be seen that their main shapes are almost the same, but the result by the WT method is superior to that of the conventional method in both shape and noise level in the high-k region.

Method B can also be used for removing the background in this example. Figure 5.26 shows the decomposed approximations obtained by using

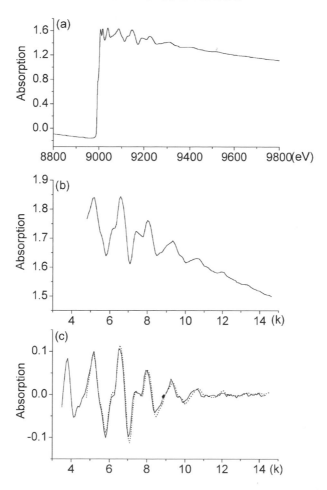

Figure 5.24. An experimental EXAFS spectrum of Cu sample (a), the converted spectrum in k space (b), and background-removed results obtained by conventional (c-solid) and WT (c-dot) methods.

Equations (5.32) and (5.33) of the improved algorithm. It can be seen that the oscillation signal is gradually removed from c_0 to c_4, but the smoothed background remains. Therefore, if we subtract c_4 from c_0, the oscillation part can be obtained, which is shown in Figure 5.27, curve (a) by the dotted line. The solid line shows the results obtained by the conventional cubic spline interpolation method for comparison. Comparing with the two curves in Figure 5.27, curve (a), we can also find that there is no significant difference between them except that the results obtained by the WT method are superior to those obtained by the conventional method in both shape and noise level in the high-k region.

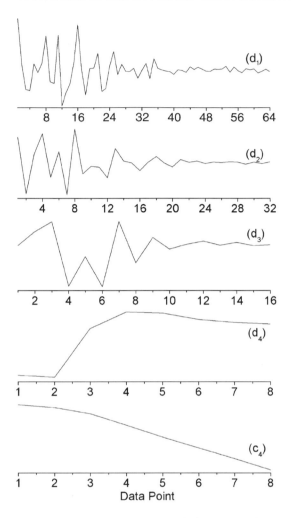

Figure 5.25. Plots of the decomposed approximation (c_4) and details (d_1, d_2, d_3, d_4) obtained by applying WT to the k-space EXAFS spectrum with the MRSD algorithm.

Computational Details of Example 5.8

Method A:

1. Load experimental spectrum (Fig. 5.24a) and the spectrum in k space, (Fig. 5.24b).
2. Extend the spectral data to an integer power of 2.
3. Make a wavelet filter—Daubechies4.
4. Set resolution level $J = 4$ $(8 - 4)$.

5. Perform forward WT to obtain the wavelet coefficients.
6. Display Figure 5.25.
7. Set approximation coefficients with zeros, and construct the signal by applying inverse WT.
8. Display Figure 5.24

Method B:

1. Load experimental spectrum (Fig. 5.24a).
2. Extend the spectral data for avoiding the edge effect.
3. Make a wavelet filter—Daubechies4.
4. Set resolution level $J = 5$.
5. Perform forward WT to obtain the **c** and **d** components with the improved algorithm.
6. Display Figure 5.26.
7. Subtract c_4 from the experimental spectrum.
8. Convert the subtracted result to k space.
9. Display Figure 5.27, curve (a).

As mentioned above, the aim of analyzing the EXAFS spectrum is to obtain the structural parameters such as N and r in Equation (5.47). In order to obtain the structural parameters from the EXAFS oscillation, Fourier filtering and least-square fitting can be performed. Figure 5.27, curve (b) shows the filtered results for the first coordination shell from the EXAFS signals in Figure 5.27, curve (a), and Table 5.3 compares the structural parameters obtained by least-square fitting of three Cu samples. The last column in the table shows a comparison of the fitted errors, which is

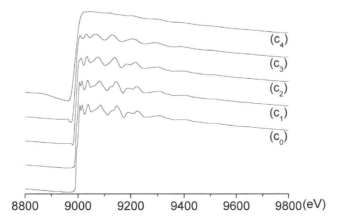

Figure 5.26. Plots of the approximations obtained by applying WT to the experimental EXAFS spectrum with the improved algorithm.

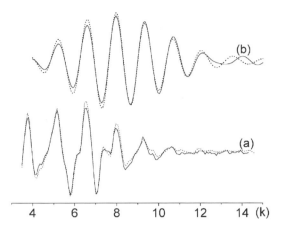

Figure 5.27. Background-removed results by conventional (solid line) and WT (dotted line) methods (a), and their filtered results (b).

calculated by

$$E = \frac{1}{N} \sum_{i=1}^{N} (X_{cal}^i - X_{exp}^i)^2 \qquad (5.48)$$

where N is the number of points in the spectra and X_{cal}^i and X_{exp}^i are, respectively, the fitted and the experimental values. For the sake of comparison between the two methods, both of the experimental and the fitted spectra were normalized when calculating the fitted error.

From Figure 5.27, curve (b), it is clear that the result by WT method is superior to that of the cubic spline method. From Table 5.3 it can be seen that, except for the coordination distance r, which is very close to the results of the two methods, all the other three parameters obtained by the wavelet transform method are larger than the results of cubic spline method. But they are reasonable. The fitted errors are also improved by the WT method. Table 5.3 also shows that the reproducibility of the three results

Table 5.3. Comparison of Structural Parameters and Fitted Errors Obtained by Least-Squares-Fitting from Background-Removed Spectra with WT and Spline Method Respectively

Spectrum	Method	N	r	σ	λ	Fitted Error
I	WT	12	2.50	0.112	6.5	0.0026
	Spline	8	2.52	0.077	4.4	0.0074
II	WT	12	2.49	0.109	6.3	0.0029
	Spline	9	2.53	0.082	4.5	0.0027
III	WT	12	2.51	0.113	6.3	0.0011
	Spline	9	2.52	0.084	4.5	0.0056

obtained by wavelet transform method is obviously superior to that of the cubic spline method. The reason for this is that the cubic spline method performs the background removal according to the points selected by the operator. However, there is no operator interference in the WT method.

5.3.3. Baseline Correction

Baseline drift is caused mainly by continuous variations of experiment conditions, such as temperature, solvent programming in liquid chromatography, or temperature programming in gas chromatography. Therefore, baseline drift is a very common problem in chromatographic studies.

Figure 5.28 shows an example of the separation of the drifting baseline from a chromatogram with gradient elution by method B. Curve (a) is the experimental chromatogram. From the figure, it can be seen that there is a strong baseline drift caused by the gradient elution in the chromatogram. Curve (b) is the 8th-scale discrete approximation c_8 decomposed by WT with Symmlet (S5) wavelet. Apparently, it resembles the baseline. Figure 5.29 shows the result obtained by subtracting curve (b) from curve (a) of Figure 5.28 with a factor f of 0.93. From the result, it is clear that the removal of baseline by this method is complete and satisfactory.

5.3.4. Background Removal Using Continuous Wavelet Transform

As stated in Chapter 4, the CWT of a signal $s(t)$ with an analyzing wavelet $\psi(t)$ is the convolution of $s(t)$ with a scaled and conjugated wavelet

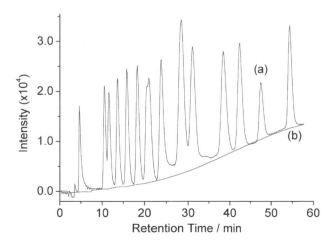

Figure 5.28. An experimental chromatogram (a) and its eighth discrete approximation obtained by WT decomposition (b).

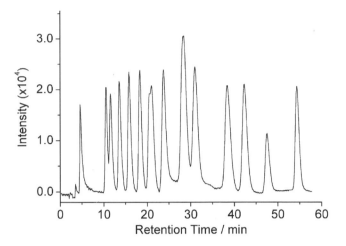

Figure 5.29. Baseline-corrected chromatogram obtained by subtracting the eigth discrete approximation from the experimental chromatogram.

$\psi_a(t) = \overline{\psi_a(-t)}$:

$$Wf(a, b) = \psi_a * s(b) = \frac{1}{\sqrt{|a|}} \int_{-\infty}^{+\infty} \overline{\psi\left(\frac{t-b}{a}\right)} f(t) dt \qquad (5.49)$$

In Fourier domain, the equation takes the form

$$Wf(a, b) = \frac{1}{2\pi} \int_{-\infty}^{+\infty} \overline{\hat{\psi}(a\omega)} \hat{s}(\omega) e^{i\omega b} d\omega \qquad (5.50)$$

where $\hat{\psi}$ and $\hat{s}$ are the Fourier transforms of the wavelet ψ and the signal s, respectively. Equations (5.49) and (5.50) show clearly that the wavelet analysis is a time–frequency analysis, or, more properly, a timescale analysis because the scale parameter a behaves as the inverse of a frequency. In particular, Equation (5.50) shows that the CWT of a signal is a filter with a constant relative bandwidth $\Delta\omega/\omega$. Therefore, the CWT should be used for separating the smooth background and the sharp peaks. In the following paragraphs, a method for removal of large spectral line from NMR spectrum is introduced.

Let $s(t)$ be a signal of the form

$$s(t) = \sum_{l=1}^{N} s_l(t) \qquad (5.51)$$

where $s_l(t) = A_l(t) \exp(i\omega_l t)$ is the lth spectral line, which has a constant frequency $f_l = \omega_l/2\pi$, and N is the number of the spectral lines. Its CWT is given by

$$Wf(a, b) = \sum_{l=1}^{N} Wf_l(a, b) \tag{5.52}$$

and the Wf_l is

$$
\begin{aligned}
Wf_l(a, b) &= \frac{1}{2\pi} \int_{-\infty}^{+\infty} \hat{\psi}(a\omega)\hat{A}_l(\omega - \omega_l)e^{i\omega b} d\omega \\
&= \frac{1}{2\pi} e^{i\omega_l b} \int_{-\infty}^{+\infty} \hat{\psi}(a(\omega + \omega_l))\hat{A}_l(\omega)e^{i\omega b} d\omega
\end{aligned}
\tag{5.53}
$$

Using the Taylor expansion of the Fourier transform of the analyzing wavelet $\hat{\psi}$ around the pulsation ω_l

$$\hat{\psi}(a(\omega + \omega_l)) = \hat{\psi}(a\omega_l) + \sum_k \frac{(a\omega)^k}{k!} \frac{d^k\hat{\psi}}{d\omega^k}(a\omega_l) \tag{5.54}$$

we obtain the following expansion for Wf_l:

$$Wf_l(a, b) = \hat{\psi}(a\omega)s_l(b) + e^{i\omega_l b} \sum_{k \geq 1} \frac{(-ia)^k}{k!} \frac{d^k\hat{\psi}}{d\omega^k}(a\omega_l)\frac{d^k A_l}{db^k}(b) \tag{5.55}$$

Therefore, we have

$$Wf_l(a, b) \approx \hat{\psi}(a\omega_l)s_l(b) \tag{5.56}$$

and

$$Wf(a, b) \approx \sum_{l=1}^{N} \hat{\psi}(a\omega_l)s_l(b) \tag{5.57}$$

If the values of frequency ω_l are sufficiently far away from each other, the factor $\hat{\psi}(a\omega)$ will allow us to treat each spectral line independently. In this case, the contribution of the lth spectral line to the $Wf(a, b)$ is localized on the scale $a_l = \omega_0/\omega_l$, where ω_0 is the frequency of the analyzing wavelet. Therefore, we have

$$\frac{Wf(\omega_0/\omega_l, b)}{\hat{\psi}(\omega_0)} \approx s_l(b) \tag{5.58}$$

Using this equation, we can easily separate the large spectral line and small peaks. However, in many cases, especially when the frequency of the each component are close to each other, we cannot obtain satisfactory results using this equation because

$$\frac{Wf(a_l, b)}{\hat{\psi}(\omega_0)} \approx s_l(b) + \Delta Wf(a_l, b) \tag{5.59}$$

the second term in the equation is a sum over the other spectral lines, with the amplitudes attenuated by the exponential factor.

Therefore, in practical applications, we can define

$$s_l^{(k)}(b) = \frac{Wf(a_l, b)}{\hat{\psi}(\omega_0)} \tag{5.60}$$

and iterate the procedure with $s_l^{(k-1)}(b)$ as the new input signal. After a certain number of iterations, the second term in Equation (5.59) will become negligible.

Example 5.9: Large Spectral Line Removal of a Simulated NMR Spectrum. The NMR signal in Fourier domain can be simulated by

$$s_l(t) = A_l(t) \exp(i\omega_l t) = A_l \exp(-d_l t) \exp(i\omega_l t) \tag{5.61}$$

Figure 5.30 shows three simulated signals $s_1(t)$, $s_2(t)$, and $s(t) = s_1(t) + s_2(t)$ and their Fourier transforms, where $s_1(t)$ and $s_2(t)$ were simulated by

$$s_1(t) = 1.0 \exp\frac{-t}{200} \exp(i0.2\pi t) \tag{5.62}$$

$$s_2(t) = 10.0 \exp\frac{-t}{100} \exp(i0.19\pi t) \tag{5.63}$$

where t is sampled by 2048 data points. It can be seen that the large spectral line $s_1(t)$ can be viewed as the baseline or the background of the small peak $s_2(t)$.

In order to separate the signals $s_1(t)$ and $s_2(t)$ from the mixed signal, we can use the iterative procedure described by Equation (5.60). Figure 5.31 shows the results at the number of the iteration $k = 100$, 200, and 300. In the calculation, Morlet wavelet, which is defined by

$$\psi(t) = e^{i\omega_0 t} e^{-t^2/2\sigma_0^2} \tag{5.64}$$

$$\hat{\psi}(\omega) = \sqrt{2\pi}\sigma_0 e^{-(\omega-\omega_0)^2\sigma_0^2/2} \tag{5.65}$$

and $\sigma_0 = 1$, $\omega_0 = 5$ were adopted. Because the aim is to remove the $s_1(t)$ and extract $s_2(t)$, $a_l = \omega_0/\omega_1 = 5/0.2\pi$ was used. It can be seen that, after 500 iterations, the large spectral line is completely removed.

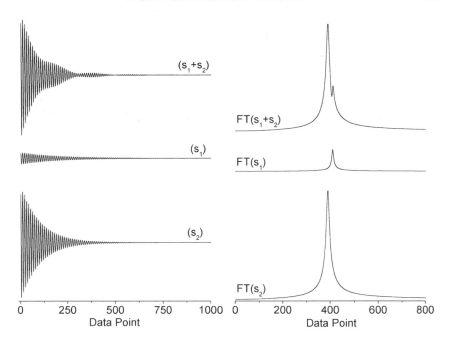

Figure 5.30. Simulated NMR signals in Fourier domain (left) and the plots of their Fourier transform (right).

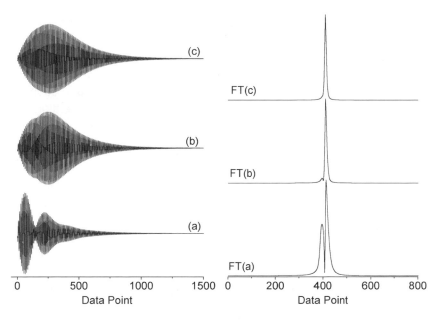

Figure 5.31. Plots of the extracted NMR signals (left) and their Fourier transform at $k = 100$ (a), 300 (b), and 500 (c) (right).

Computational Details of Example 5.9

1. Generate the mixed signal in Figure 5.30 with $\omega_1 = 0.2\pi$ and $\omega_2 = 0.19\pi$.
2. Set the scale parameter for the CWT: $a = \omega_0/\omega_1$, where the ω_0 is 5, which is determined by wavelet function used in the CWT.
3. Perform CWT to remove the large spectral line iteratively.
4. Display Figure 5.31.

Using the same procedure, D. Barache et al. successfully subtracted a large component from an experimental NMR spectrum of polyethylene as shown in Figure 5.32a. The huge line is the peak corresponding to CH_2 groups, which completely obliterates the fine details of the other peaks. After subtraction of the large peak as shown in Figure 5.32b, the small peaks become clearly identifiable.

5.3.5. Background Removal of Two-Dimensional Signals

With the development of modern instruments, more and more analytical instruments provide two-way data matrices as their measurement results. A method for background removal of 2D analytical signals was also proposed on the basis of the WT technique.

Assume a data matrix **X** of order $n \times m$, in which each line represents a spectral measurement with m wavelength sampling points and each column represents a chromatographic measurement with n retention-time

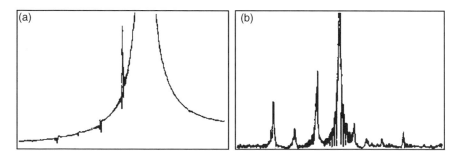

Figure 5.32. An experimental NMR spectrum with large spectral line (a) and the spectrum obtained by CWT extraction (b) [copied from D. Barache et al. *J. Magn. Reson.* **128**:1–11, (1997)].

sampling points. The data matrix can be divided into two parts:

$$\mathbf{X} = \mathbf{X}_c + \mathbf{X}_b \tag{5.66}$$

Here $\mathbf{X}_c$ originated from chemical components and $\mathbf{X}_b$ from noise and background. For the sake of convenience, we assume that the data matrix is free of noise; $\mathbf{X}_b$ will composed of spectral and chromatographic background. The background of 2D analytical data matrices such as the "hyphenated" chromatographic–spectroscopic data generally has the following properties: (1) there is no direct correlation between the chromatographic baseline drift and the spectral background, (2) there is a very similar spectral background at the two ends of a chromatographic peak, and (3) there is a similar drift of the baseline at each retention time since the scanning time for each spectrum is very short. Thus, the background matrix can be written as

$$\mathbf{X}_b = \mathbf{t}\mathbf{1}^T + \mathbf{1}\mathbf{s}^T \tag{5.67}$$

where $\mathbf{t}$ denotes the baseline drift in chromatographic direction, $\mathbf{s}^T$ denotes the spectral background, and $\mathbf{1}$ and $\mathbf{1}^T$ denote that a vector contains only 1s. The superscript T denotes transposition.

Therefore, the spectrum at retention time i can be expressed as

$$\mathbf{x}_i^T = \mathbf{x}_{c,i}^T + t_i \mathbf{1}^T + \mathbf{s}^T \tag{5.68}$$

where $\mathbf{x}_{c,i}^T$ denotes a "pure" spectrum at retention time i and t_i corresponding to the baseline drift at the retention time. In a zero-component regions, $\mathbf{x}_{c,i}^T$ should be a zero vector. Equation (5.68) turns into

$$\mathbf{x}_i^T = t_i \mathbf{1}^T + \mathbf{s}^T \tag{5.69}$$

Equation (5.69) shows that the spectra in zero-component regions before and after elution should be similar if there is spectral background. If not, they should be flat lines. Therefore, the zero component regions may be used to detect the presence of spectral background.

According to the algorithm of WT, we can obtain approximation and detail coefficients by using Equations (5.3) and (5.4). The detail coefficients of a spectrum at retention time i on scale k can be expressed as

$$
\begin{aligned}
\mathbf{d}_k &= \mathbf{x}_i^T \mathbf{H}_0 \mathbf{H}_1 \mathbf{H}_2 \cdots \mathbf{G}_{k-1} \\
&= \mathbf{x}_{c,i}^T \mathbf{H}_0 \mathbf{H}_1 \mathbf{H}_2 \cdots \mathbf{G}_{k-1} + t_i \mathbf{1}^T \mathbf{H}_0 \mathbf{H}_1 \mathbf{H}_2 \cdots \mathbf{G}_{k-1} \\
&\quad + \mathbf{s}^T \mathbf{H}_0 \mathbf{H}_1 \mathbf{H}_2 \cdots \mathbf{G}_{k-1}
\end{aligned} \tag{5.70}
$$

According to the properties of the filters $\mathbf{H}$ and $\mathbf{G}$, it is easy to deduce that the detail coefficients of a constant vector $\mathbf{c} = \{c, c, \ldots, c\}$ should be a zero vector:

$$\mathbf{d}_1 = \mathbf{cG} = \mathbf{0} \tag{5.71}$$

Using this property, Equation (5.70) can be reduced to

$$\mathbf{d}_k = \mathbf{x}_{c,i}^T \mathbf{H}_0 \mathbf{H}_1 \mathbf{H}_2 \cdots \mathbf{G}_{k-1} + \mathbf{s}^T \mathbf{H}_0 \mathbf{H}_1 \mathbf{H}_2 \cdots \mathbf{G}_{k-1} \tag{5.72}$$

In zero-component regions, this last equation can be further reduced into

$$\mathbf{d}_k = \mathbf{s}^T \mathbf{H}_0 \mathbf{H}_1 \mathbf{H}_2 \cdots \mathbf{G}_{k-1} \tag{5.73}$$

Therefore, using Equations (5.72) and (5.73), we can remove both the chromatographic baseline drift and the spectral background.

For example, Figure 5.33a,b shows two simulated chromatographic peaks and two simulated spectra, which are used to simulate a theoretical HPLC-DAD data matrix (denoted as $\mathbf{X}$). Figure 5.33c,d shows a simulated chromatographic baseline and a simulated spectral background, which are used to simulate an HPLC-DAD data matrix containing baseline drift and background. We can simulate two data matrices $\mathbf{X}_1$ and $\mathbf{X}_2$,

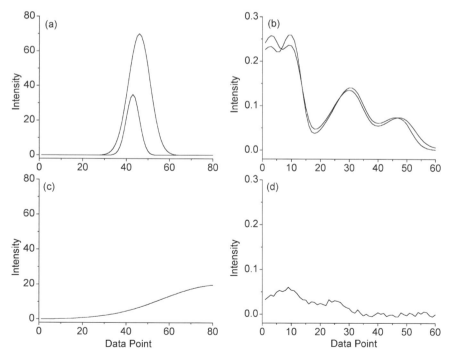

Figure 5.33. Chromatograms (a), spectra (b), chromatographic baseline (c), and spectral background (d) used in the simulation of the HPLC-DAD data matrices.

where X_1 includes only chromatographic baseline drift and X_2 includes both chromatographic baseline drift and spectral background.

In order to remove the baseline drift and the spectral background from the data matrices X_1 and X_2, we can apply a WT decomposition to each line of the two data matrices and use the detail coefficients on a definite scale k to combine new data matrices X_{1d} and X_{2d}. In both matrices X_{1d} and X_{2d}, chromatographic baseline should have been corrected. After that, for the matrix X_{2d}, we can further remove the spectral background by subtracting the transformed background from it using the information in zero-component regions. Note that in practice, it is very important to determine the decomposition scale k. Generally, $k = 1$ or 2 can be used for those signals with a low sampling frequency, but $k = 3$, 4, or 5 should be used for those signals with a high sampling frequency. In this example, $k = 2$ was used. Furthermore, different wavelet basis used give different results. There is no general rule for selection of the best wavelet basis for a given signal except trial and error. In this example, Daubechies wavelet filters with length $L = 4$ were used.

In order to investigate the effect of the method, we can examine the rankmap of the data matrices as it can reveal the factor number of these data matrices. Figure 5.34 shows the rankmaps of the data matrices X, X_1, X_2 before and after background removal. By comparing the rankmaps before and after background removal, it is clear that the factor corresponding to the chromatographic baseline and the spectral background in panels (a), (b), and (c) disappeared in panels (d), (e) and (f) (of Fig. 5.34). This indicates that both the chromatographic baseline and the spectral background are completely removed.

An experimental HPLC-DAD data matrix was processed by the method in an article published in *Chemometrics and Intelligent Laboratory Systems* [**37**:261–269 (1997)] by Shen et al. Figure 5.35a shows the rankmap of the data matrix of a sample consisting of two isomers of a pharmacologically active drug. It is clear that the factor number is not correct. There are two factors caused by background absorption. But in Figure 5.35b, which shows the rankmap of the matrix after background removal by the method, only two main factors remain. In this work, decomposition scale $k = 3$ and Daubechies wavelet filters with length $L = 4$ were adopted.

5.4. RESOLUTION ENHANCEMENT

Poor resolution is a very common problem encountered by chemists in spectral or chromatographic studies. In spectral studies, resolution of a spectrum is very important to ensure a correct assignment of the spectral

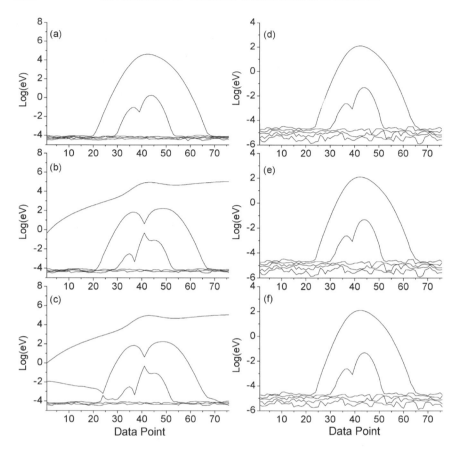

Figure 5.34. Rankmaps of the data matrices $\mathbf{x}, \mathbf{x}_1, \mathbf{x}_2$ before (a),(b),(c) and after (d),(e),(f) background removal by WT decomposition.

peaks. The problem of poor resolution is always an obstruction in spectral analysis. Another problem in both spectral and chromatographic studies is peak overlapping. We must resort to mathematical techniques to solve these problems. Usually, these problems are solved by using techniques such as linear or nonlinear regression analysis, curve-fitting procedures, derivative procedures, neural networks, and chemical factor analysis.

5.4.1. Numerical Differentiation Using Discrete Wavelet Transform

Derivative calculation is a powerful technique used in analytical chemistry to resolve spectra, sharpen peaks, determine potentiometric titration end-points, carry out quantitative analysis, and perform similar procedures.

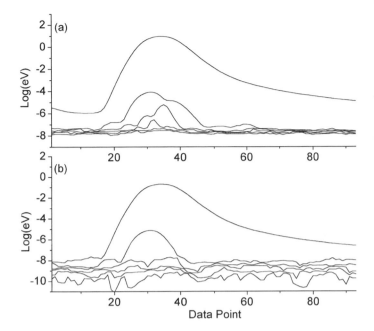

Figure 5.35. Rankmap of an experimental HPLC-DAD data matrix before (a) and after (b) background removal by WT decomposition.

Although derivative calculation is a useful tool for data analysis, it has a major drawback in increasing the noise level in higher-order derivative calculation. In order to improve the signal-to-noise ratio (SNR) of higher-order derivative calculation via conventional numerical differentiation, noise reduction is usually performed before calculating the successive order derivative, which will lead to inconvenience and complication in the calculation.

In practice, the simplest method of derivative calculation is numerical differentiation. Other methods used in chemical studies for derivative calculation include Fourier transform and the Savitsky–Golay method. The former method cannot improve the SNR in its result, and the latter method involves the selection of suitable parameters.

On the basis of the characteristics of the wavelet functions, a method for derivative calculation using DWT with Daubechies wavelet was proposed. The approximate first derivative of a signal $\mathbf{x}$ can be computed by the difference between two scale coefficients

$$\mathbf{x}^{(1)} = \mathbf{c}_{j,D_{2m}} - \mathbf{c}_{j,D_{2\tilde{m}}} \qquad m \neq \tilde{m} \qquad (5.74)$$

where $\mathbf{x}^{(1)}$ represents the first derivative of the signal $\mathbf{x}$, D_{2m} and $D_{2\tilde{m}}$ represent two Daubechies wavelet functions, and $\mathbf{c}_{j,D_{2m}}$ and $\mathbf{c}_{j,D_{2\tilde{m}}}$ are the

jth-scale discrete approximations obtained by Equations (5.1) and (5.2) with the two wavelets, respectively. In practice applications, $j = 1$ is generally used. But for noisy signals, a higher value of j can be used to improve the SNR of the calculated result.

Higher-order derivative computation can be achieved by using the result obtained from the lower-derivative calculation as an input for WT calculation

$$\mathbf{x}^{(n)} = \mathbf{c}_{j,D_{2m}}^{(n-1)} - \mathbf{c}_{j,D_{2\tilde{m}}}^{(n-1)} \quad m \neq \tilde{m} \text{ and } n > 1 \tag{5.75}$$

where $\mathbf{x}^{(n)}$ represents the nth-order derivative and $\mathbf{c}_{j,D_{2m}}^{(n-1)}$ and $\mathbf{c}_{j,D_{2\tilde{m}}}^{(n-1)}$ represent the jth-scale approximation coefficients obtained by WT from the $(n - 1)$th-order derivative with the Daubechies wavelet functions D_{2m} and $D_{2\tilde{m}}$.

Example 5.10: Approximate Derivative Calculation of Simulated Signals Using DWT. Figure 5.36a shows three types of typical analytical

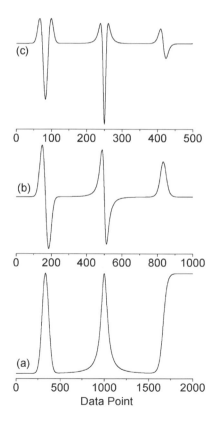

Figure 5.36. A simulated signal (a) and its first- (a) and second- (c) order derivatives calculated by DWT method.

signals simulated by Gaussian, Lorentzian, and sigmoid functions

$$x_{gauss} = A_0 \exp\left(-4\ln(2)\left(\frac{t - t_0}{W_{1/2}}\right)^2\right) \tag{5.76}$$

$$x_{lorentz} = A_0\left(1 + 4\left(\frac{t - t_0}{W_{1/2}}\right)^2\right)^{-1} \tag{5.77}$$

$$x_{sigmoid} = \frac{A_0}{1 + e^{-k(t-t_0)}} \tag{5.78}$$

where A_0, t_0, and $W_{1/2}$. are the amplitude, position, and the full width at half maximum of the simulated peak, respectively, and k is a parameter to control the gradient of the curve. Figure 5.37a shows a noisy signal by the curve (a) in Figure 36 plus a random noise at SNR $= 100$.

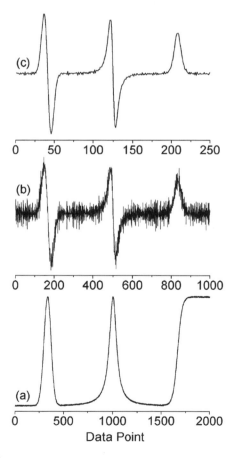

Figure 5.37. A simulated noisy signal (a) and its first derivative calculated by DWT with $j = 1$ (b) and $j = 3$ (c), respectively.

Figure 5.36b,c shows the first- and second-order derivatives obtained by Equations (5.74) and (5.75), respectively, using $j = 1$ and Daubechies $N = 18$ and 8 as the filters (D_{2m} and $D_{2\tilde{m}}$). It can be seen that very good results can be obtained. The only drawback is that the number of data points is reduced to $\frac{1}{2}$ for the first-order derivative and $\frac{1}{4}$ for the second-order derivative.

Figure 5.37b shows the first derivative using the same method and the parameters as in Figure 5.36b. It can be seen that, when the signal is noisy, the SNR of the result will be even worse than that of the signal. In this case, we can use a higher value of j for the calculation. Figure 5.37c is obtained using $j = 3$, which is an acceptable result. But there is also the problem of the reduction of the data point number. The number of data point in Figure 5.37c is only $\frac{1}{8}$th of the original length.

Although interpolation can solve the problem of the data point number, we can also use Equations (5.32) and (5.33) for calculation of the $c_{j,D_{2m}}$ and $c_{j,D_{2\tilde{m}}}$ in Equation (5.74) or $c_{j,D_{2m}}^{(n-1)}$ and $c_{j,D_{2\tilde{m}}}^{(n-1)}$ in Equation (5.75). The benefit of using the two equations is that the number of data points in c_j does not change. Consequently, the length of the calculated derivatives will remain the same as that of the original signal. Figures 5.38 and 5.39 show the first- and the second-order derivatives of the simulated signals with the same filters as in the DWT method and $j = 1$ and $j = 4$, respectively. It can be seen that all the results are acceptable. Only the symmetry of the peaks is slightly distorted because of the effect of the noise in Figure 5.39c.

Computational Details of Example 5.10

1. Generate the signal with 2000 data points (Fig. 5.36a) by using Gaussian, Lorentzian, and sigmoid equations.
2. Extend the data point number to 2048.
3. Make two wavelet filters—Daubechies18 and 8.
4. Set resolution level $J = 1$.
5. Perform DWT with the two filters, respectively, to obtain the wavelet coefficients.
6. Calculate the first derivative by subtracting the approximate coefficients one from another.
7. Display Figure 5.36b.
8. Perform DWT on the coefficients obtained in step 5 with the two filters, respectively.
9. Calculate the second derivative from the approximate coefficients.
10. Display Figure 5.36c.

Note: The results in Figure 5.37 were obtained in a similar way with the only difference in $J = 3$ for the second-derivative calculation, and the results in Figures 5.38 and 5.39 were obtained by using the improved algorithm with $J = 1$ and 4, respectively.

5.4.2. Numerical Differentiation Using Continuous Wavelet Transform

CWT with some specific wavelet functions can also be used for approximate derivative calculation of analytical signals. The Haar wavelet function, for instance, is one of the appropriate wavelet bases because of its symmetric

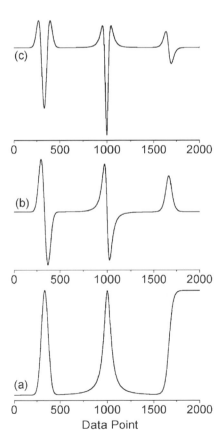

Figure 5.38. A simulated signal (a) and its first- (b) and second- (c) order derivatives calculated by the improved WT algorithm with $j = 1$.

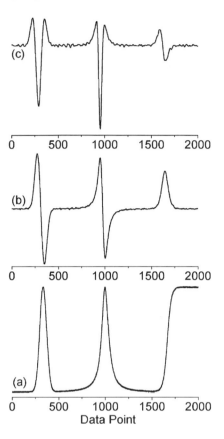

Figure 5.39. A simulated noisy signal (a) and its first- (b) and second- (c) order derivatives calculated by the improved WT algorithm with $j = 4$.

property. Just as in the method using DWT, the approximate nth-order derivative of an analytical signal can be obtained by applying n times of the CWT to the signal. CWT is continuous in terms of scaling and shifting; that is, during computation, the analyzing wavelet can be shifted smoothly over the full domain of the analyzed signal at any scale. Therefore, it possesses much stronger resolving ability than does DWT in both time and frequency domains. In comparison with the other methods, the SNR of the derivatives for noisy signals can be greatly improved using the CWT method with a proper scale parameter a for $\psi_a(t)$ in Equation (5.49).

In practice, analytical signals are discrete; thus, a discrete form of the Equation (5.49) is used

$$Wf(a, i\Delta_s) = \frac{\Delta_s}{\sqrt{|a|}} \sum_n f(n\Delta_s)\psi\left(\frac{(n-i)\Delta_s}{a}\right) \tag{5.79}$$

where i and n are indices of the data point, Δ_s corresponds to the sampling interval, and $\psi(t)$ is the Haar wavelet function:

$$\psi(t) = \begin{cases} 1 & 0 \le t < \frac{1}{2} \\ -1 & \frac{1}{2} \le t < 1 \\ 0 & \text{other} \end{cases} \tag{5.80}$$

It is obvious that because the Haar wavelet function is of a ladder shape, the result from Equation (5.79) should correspond to the first derivative of $f(n\Delta_s)$, and the nth-order derivative can be obtained by applying CWT to the $(n-1)$th-order derivative.

Example 5.11: Approximate Derivative Calculation of Simulated Signals Using CWT. Figures 5.40 and 5.41 show the first- and second-order derivatives of the two simulated signals used in the DWT method above. They are computed by using the CWT method with different values of the scale parameter a. It can be seen that, for the signal without noise in Figure 5.40, all the results at different values of a are satisfactory. Only the width of the peaks in the calculated derivatives becomes increasingly broad with increase in the value of a. But for the noisy signal in Figure 5.41, it can be seen that the effect of noise can be eliminated with a relative large value of a.

Computational Details of Example 5.11

1. Generate the signal with 2000 data points (Fig. 5.36a) using Gaussian, Lorentzian, and sigmoid equations.
2. Extend the data point number to 2048.
3. Perform CWT with Haar wavelet and the scale parameter $a = 1$, 20, and 50 to obtain the first- and second-order derivatives, and display them in Figure 5.40 and 5.41, respectively, for the simulated signal with and without noise.

Taking the Gaussian peak as an example, we can investigate the effect of the scale parameter a on the peak width of its derivative. Figure 5.42 shows the relationship between increase in peak width and the value of a. The increase is represented in percent of the peak width in the derivative by the conventional numerical method. The values of the circled points connected by the dotted line are obtained by manual measurement, and the solid line represents the fitted curves. It is clear that the peak width

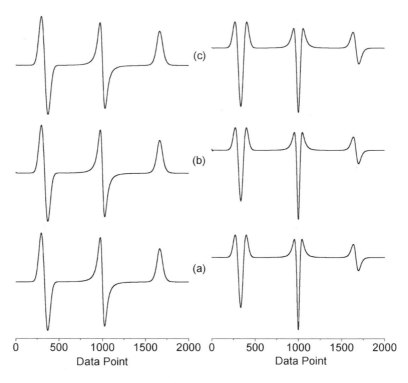

Figure 5.40. The first- (left column) and second- (right column) order derivatives of the simulated signal without noise calculated by CWT with Haar wavelet and the scale parameter $a = 1$ (a), 20 (b), and 50 (c).

increases with the value of a, but the increase is less than 10% when the parameter a is less than 100.

Figure 5.43, curve (a) shows a photoacoustic spectrum of a rare-earth compound. There are three overlapping absorption peaks in the range of 400–500 nm. It is evident that high magnitude noise exists in the spectrum, which makes the position of peaks hard to determine. If the spectral data are directly processed by numerical differentiation methods, the SNR of the derivatives will be too low to see anything but noise. Figure 5.43, curves (b) and (c) are respectively the first and second derivatives computed by the CWT method with $a = 100$. It can be seen that the derivatives are smooth and clean, and the SNR is even better than that of the original signal. Although there is still a small-magnitude noise in the derivatives, it does not affect determination of the position of peaks, because the overlapping peaks are clearly resolved in the derivatives.

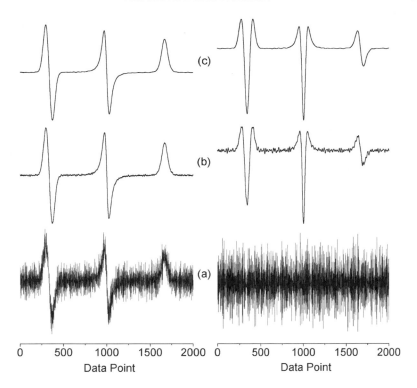

Figure 5.41. The first- (left column) and second- (right column) order derivatives of the simulated noisy signal calculated by CWT with the Haar wavelet and the scale parameter $a = 1$ (a), 20 (b), and 50 (c).

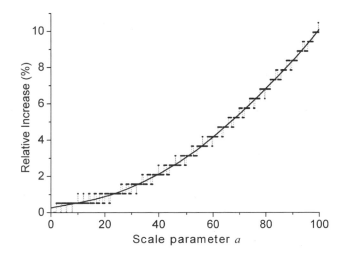

Figure 5.42. Relationship between the relative increase of the peak width in the calculated derivative of a Gaussian peak and the value of the parameter a.

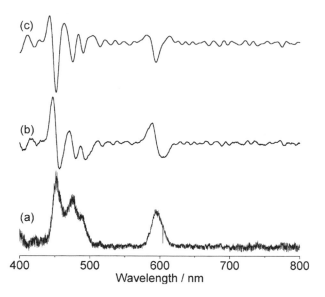

Figure 5.43. An experimental photoacoustic spectrum (a) and its first- (b) and second-(c) order derivatives calculated by CWT with $a = 100$ [From *Fresenius J. Anal. Chem.* **367**:525–529 (2000)].

5.4.3.   Comparison between Wavelet Transform and other Numerical Differentiation Methods

Here, we compared the derivatives obtained using different methods, including numerical differentiation, the Fourier transform method, the Savitsky–Golay method, the DWT method, the DWT method with the improved WT algorithm, and the CWT method with $a = 50$. The first-order derivatives of the above two simulated signals by these methods are shown in Figure 5.44, respectively. From the figure it can be seen that, for the signal without noise, all the results are almost the same. But for the noisy signal, only the last two methods can obtain smooth derivatives.

Table 5.4 lists the parameters describing these results in the left column of Figure 5.44. The definition of the peak width and peak position is depicted in Figure 5.44, curve (f). From the parameters in the table, we can find that all the methods give the same peak position, but the peak width is slightly different. Because it is difficult to determine the exact beginning and ending points of the peaks, there is a large error in the peak width in the table. But it is evident that the peak width in the results using the CWT method is comparatively greater than from the other methods.

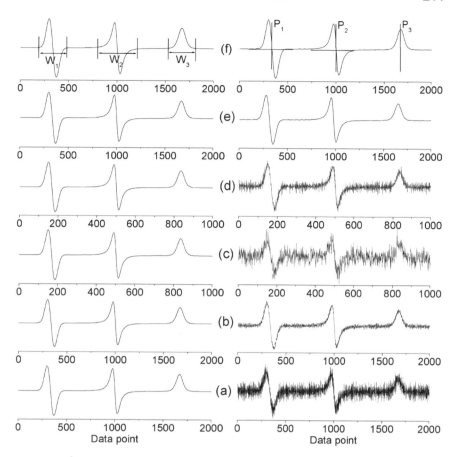

Figure 5.44. Comparison of the first derivatives of the clean (left column) and noisy (right column) simulated signals calculated by numerical differentiation (a), Fourier transform method (b), Savitsky–Golay method (c), DWT method (d), DWT method with the improved WT algorithm (e), and the CWT method (f).

Table 5.4. Comparison of Peak Position and Width of Calculated Derivatives by Different Derivative Calculation Methods

Method	Gaussian peak Position	Gaussian peak Width	Lorentzian peak Position	Lorentzian peak Width	Sigmoid peak Position	Sigmoid peak Width
Simulated	334	239	1000	596	1666	253
Numerical	333	269	1000	381	1667	215
FT	334	272	1000	370	1667	270
Savitsky–Golay	168 × 2	135 × 2	500 × 2	199 × 2	834 × 2	125 × 2
DWT	167 × 2	135 × 2	500 × 2	200 × 2	833 × 2	113 × 2
Improved DWT	333	275	1000	381	1665	241
CWT	334	330	1000	415	1667	318

5.4.4. Resolution Enhancement

According to the characteristics of the WT, a signal can be decomposed into its contributions by the MRSD method to obtain discrete details d_j and discrete approximations c_j, which represent the different components of the signal at different frequencies. If an overlapping or low-resolution signal is decomposed into its contributions, there must be discrete details that represent the information at the frequency lower than noise and higher than the original signal. Therefore, it is not difficult to select a detail at medium scale to obtain the high-resolution information of the signal. Furthermore, if we amplify one or several of these selected details and then perform the inverse transform, that is, reconstruct the original signal with the amplified contributions, we can also obtain a signal with high resolution.

Therefore, there are generally two methods using WT for resolution enhancement of analytical signals:

1. Decompose the analyzing signal c_0 into its approximations c_j and details d_j using Equations (5.32) and (5.33).
2. Inspect the c_j and d_j, then select either (a) a detail component as the resolved result for further studies (method A), or (b) one or more detail components that represent the high-resolution information of the analyzing signal (method B).
3. Multiply the selected d_j by a factor k whose value is bigger than 1.0 and reconstruct the signal by using Equation (5.34) (method B).

Example 5.12: Resolution Enhancement of an Overlapping Chromatogram Using Method A. Curve (c_0) in Figure 5.45 is an experimental chromatogram, and curves (d_4), (d_5), and (d_6) are its detail components obtained by using the improved WT algorithm in Equations (5.32) and (5.33) with Symmlet ($L = 4$) wavelet at scale parameter $j = 4$, 5, and 6, respectively.

As discussed in the WT denoising section, the frequency of the detail components d_j obtained by a WT decomposition decreases with the increase of the scale parameter j. The detail components at low scales are generally composed of noise. From Figure 5.45, it is clear that d_4 is still composed mainly of noise, but d_5 and d_6 are composed of chromatographic information whose frequency is higher than that of noise and lower than that of the original signal. That is to say, the d_5 and d_6 are the desired high-resolution part of the analytical signal. Therefore, in the detail components on medium scale, we can find one or more components which represent the high-resolution information of the analytical signal. If we further compare d_5 and d_6, it is easy to find that d_5 is the better one for representing

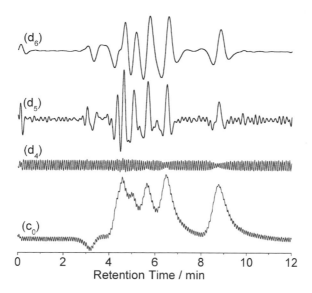

Figure 5.45. An experimental chromatogram (c_0) and its detail components obtained by WT with $j = 4(d_4)$, $5(d_5)$, and $6(d_6)$.

the high-resolution information, which can be used for further studies, such as determination of peak position and quantitative calculation.

Computational Details of Example 5.12
1. Load the experimental chromatogram [Fig. 5.45 (c_0)].
2. Extend the chromatographic data for avoiding the edge effect.
3. Make a wavelet filter—Symmlet4.
4. Set resolution level $J = 6$.
5. Perform WT to obtain the **c** and **d** components with the improved algorithm.
6. Display Figure 5.45.

Because this method can extract the high-resolution information from a low-resolution or overlapping analytical signal, a method for determination of the component number in overlapping chromatograms was proposed. Figure 5.46 shows four experimental chromatograms with different intensities. It is impossible to obtain the correct component number from such chromatograms. Figure 5.47 shows the d_5 component of the four chromatograms in Figure 5.46 obtained with Symmlet wavelet filters ($L = 4$). The dotted line in the figure indicates the position of zero in the magnitude axis. From Figure 5.47, it is clear that there are five components in the chromatograms, and all four chromatograms of different magnitude give us the same result.

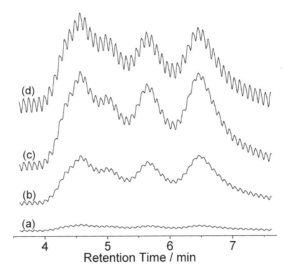

Figure 5.46. Four experimental chromatograms with different amplitudes [from *Chemometr. Intell. Lab. Syst.* **43**:147–155 (1998)].

Because the WT is a linear transform, the high-resolution information should be used for quantitative calculation. An example of quantitative determination of the components in overlapping chromatographic peaks was published in *Analytical Chemistry*, [**69**:1722–1725 (1997)]. Figure 5.48 shows the chromatograms of five mixed samples of benzene, methyl benzene, and ethyl benzene. It is difficult to perform quantitative calculation

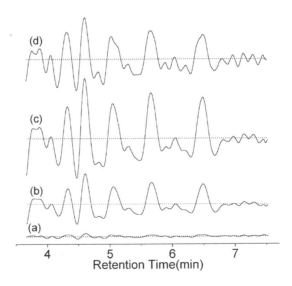

Figure 5.47. The detail component **d**$_5$ obtained from the four chromatograms in Figure 5.46 by WT decomposition [from *Chemometr. Intell. Lab. Syst.* **43**:147–155 (1998)].

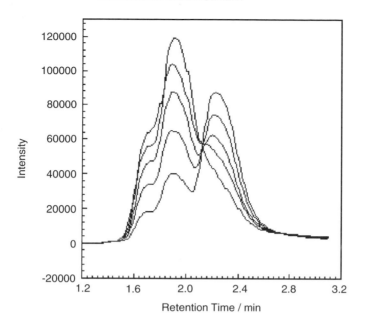

Figure 5.48. Experimental chromatograms of 5 three-component samples [from *Anal. Chem.* **69**:1722–1725 (1997)].

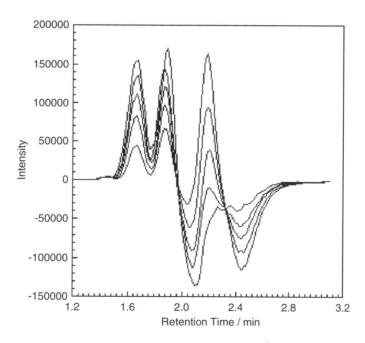

Figure 5.49. Wavelet coefficients d_3 obtained from the five chromatograms in Figure 5.48 by WT decomposition [from *Anal. Chem.* **69**:1722–1725 (1997)].

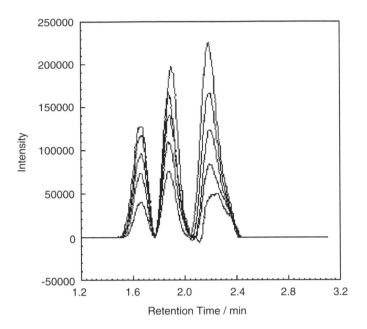

Figure 5.50. Baseline-corrected wavelet coefficients after subtracting the estimated baseline by linking the minimum point at both sides of the peak in Figure 5.49 [from *Anal. Chem.* **69**:1722–1725 (1997)].

of the components by directly using the chromatograms because of overlapping of the three peaks. Figure 5.49 shows the d_3 component obtained by WT decomposition with Haar wavelet. It is clear that the information of the three peaks is resolved. In order to calculate the peak area, we can estimate a baseline by simply linking the minimum point at both sides of a peak. Figure 5.50 shows the results after subtracting such a baseline. Figure 5.51 shows the relationship between the area and the concentration. It can be seen that a very good linearity of the signals in the wavelet coefficients is kept.

Example 5.13: Resolution Enhancement of an Overlapping NMR Spectrum Using Method B. In Figure 5.52, spectrum (a) shows a simulated NMR spectrum by the Lorentzian equation in (5.77). From left to right, the peaks are doublet, triplet, quartet, and quintet. Spectrum (b) in the figure shows the reconstructed spectrum by multiplying the d_1 and d_2 by $k_1 = k_2 = 55$. Figure 5.53 shows the detail coefficients d_1 to d_4 obtained by WT decomposition of spectrum (a) with Symmlet ($L = 4$) wavelet filters. From Figure 5.53 it can be seen that, except for discrete detail d_4, d_1 through d_3 all represent the resolved information of the peaks in the overlapping spectrum, but from d_1 to d_3 the resolution decreases. Therefore, if we amplify the details d_1 and d_2, and then perform reconstruction, the

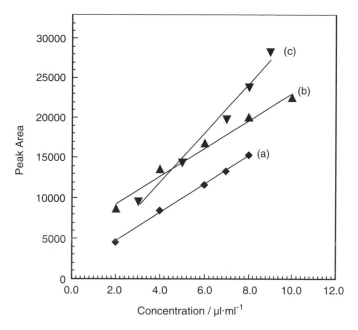

Figure 5.51. Calibration curves obtained from the peak area in Figure 5.50 and concentrations of the three components [from *Anal. Chem.* **69**:1722–1725 (1997)].

resolution of the reconstructed spectrum will be improved. From spectrum (b) in Figure 5.52, it is clear that all four groups of peak are well resolved.

Selection of the details undergoing amplification is generally by visual inspection on the decomposed details as shown in Figure 5.53. It will be

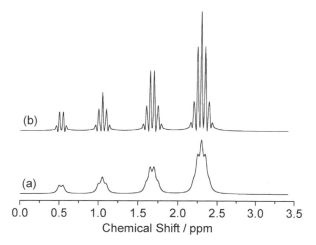

Figure 5.52. A simulated NMR spectrum (a) and the reconstructed spectrum (b) by multiplying the d_1 and d_2 by $k_1 = k_2 = 55$.

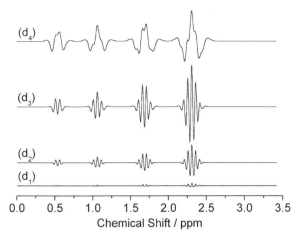

Figure 5.53. Detail coefficients of the simulated NMR spectrum obtained by WT decomposition.

difficult to select the appropriate detail coefficients when the noise level in the original signal is significant because the noise will also be decomposed into those low-scale detail coefficients for amplification. Figure 5.54, curve (a) shows a simulated noisy NMR spectrum. The detail coefficients are shown in Figure 5.55. It can be seen that the d_1 and d_2 coefficients are noisy. If we multiply d_1 and d_2 by $k_1 = k_2 = 55$ as we did above, the noise level of the reconstructed spectrum will be increased as well, as is shown in Figure 5.54, curve (b). In such cases, we can only multiply the d_2 or d_3

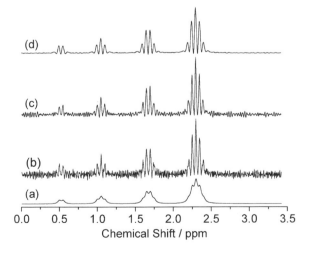

Figure 5.54. A simulated noisy NMR spectrum (a) and its reconstructed spectra by multiplying d_1 and d_2 with $k_1 = k_2 = 55$ (b), d_2 with $k_2 = 60$ (c), and d_3 with $k_3 = 10$ (d), respectively.

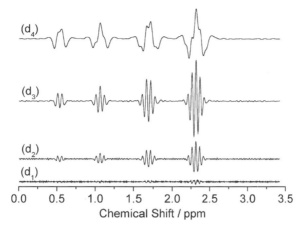

Chemical Shift / ppm

Figure 5.55. Detail coefficients of the simulated noisy NMR spectrum obtained by WT decomposition.

in order to avoid the effect of noise. Figure 5.54, curves (c) and (d) are obtained by multiplying d_2 and d_3, respectively, with $k_2 = 60$ and $k_3 = 10$. It is clear that the SNR of the results is improved.

Computational Details of Example 5.13

1. Generate the signal with 700 data points [Figure 5.52 (a)] using Lorentzian equations.
2. Make a wavelet filter—Symmlet4.
3. Set resolution level $J = 4$.
4. Perform WT to obtain the **c** and **d** components with the improved algorithm.
5. Perform reconstruction with multiplying the d_1 and d_2 with a factor 55.0.
6. Display Figures 5.52 and 5.53.

A more detailed discussion of Example 5.13 can be found in a paper published in *Applied Spectroscopy* [**54**:731–738 (2000)]. In this paper, an experimental NMR spectrum was also processed by the method. Figure 5.56 shows two enlarged parts of the experimental and the reconstructed spectra. It is clear that the resolution of the spectra is greatly improved by this method.

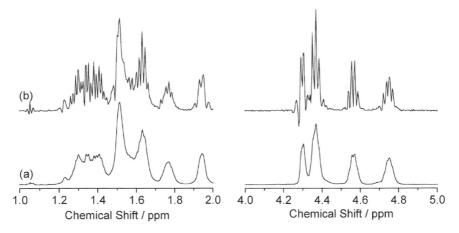

Figure 5.56. An experimental NMR spectrum (a) and the reconstructed spectrum by multiplying $\mathbf{d}_1$ and $\mathbf{d}_2$ with $k_1 = k_2 = 10$ (b).

5.4.5. Resolution Enhancement by Using Wavelet Packet Transform

WPT differs from WT with respect to the decomposition tree. In WT, only the approximation coefficients on each scale are used for further decomposition, but in WPT, the further decomposition is applied to both the approximation and detail coefficients. Therefore, for resolution enhancement, the resolving ability of WPT should be stronger than that of WT, because there will be more decomposed components representing the information with different frequencies. Consequently, it is easy for us to select a component that represents the desired high-resolution information.

The procedures for resolution enhancement of analytical signals are almost the same as in methods A and B proposed above. The only difference is to use the Equations (5.42)–(5.44) instead of Equations (5.32)–(5.34), for decomposition and reconstruction computation.

Figure 5.57 shows the experimental chromatograms of six samples containing six rare-earth ions. Concentrations of the samples are listed in Table 5.5. In order to extract the chromatographic information of each component from the overlapping chromatograms in Figure 5.57, we can subject them to WPT decomposition and obtain all $\mathbf{w}_j^p$ first. Then we can select a coefficient component to represent the desired high-resolution information. Figure 5.58 shows the selected $\mathbf{w}_6^3$ coefficients of the six chromatograms. Finally, we can estimate a baseline by linking the minimum point at both sides of every peak. After subtracting the baseline, we can obtain the results shown in Figure 5.59. It can be seen that all six peaks in the six chromatograms are well resolved except for the second peak in

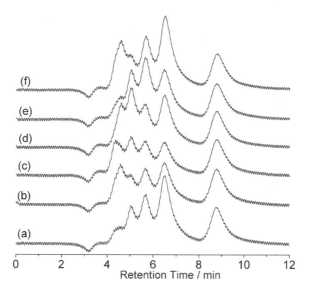

Figure 5.57. Experimental chromatograms of mixed rare-earth solution samples.

chromatogram (d), which is slightly distorted. Using the resolved peaks, we can investigate the linearity of the extracted signals by examining the relationship between the peak area and concentration. Figure 5.60 shows the calibration curves of the six components. It can be seen that all six curves are satisfactory.

5.4.6. Comparison between Wavelet Transform and Fast Fourier Transform for Resolution Enhancement

As we have seen, deconvolution by Fourier transform can also be used for resolution enhancement of analytical signals. The procedures can be

Table 5.5. Compositions and Concentrations (ppm) of Mixed Rare-Earth Solution Samples

Number	Lu	Yb	Tm	Er	Ho	Tb
1	15.0	10.0	30.0	25.0	38.0	21.2
2	25.0	30.0	15.0	18.0	22.0	21.2
3	40.0	12.0	25.0	15.0	15.0	21.2
4	16.0	38.0	40.0	18.0	26.0	21.2
5	10.0	20.0	35.0	32.0	24.0	21.2
6	34.0	34.0	16.0	30.0	35.0	21.2

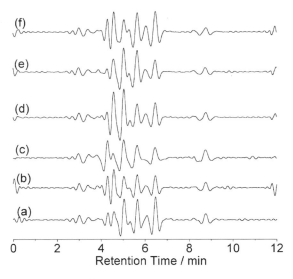

Figure 5.58. WPT coefficients $\mathbf{w}_6^3$ of the chromatograms in Figure 5.57 obtained by WPT decomposition.

summarized as follows:

1. Transform the signal $f(t)$ to the Fourier domain: $f(t) \Rightarrow \hat{f}(\omega)$.
2. Multiply $\hat{f}(\omega)$ with a window function $\hat{h}(\omega)$: $\hat{g}(\omega) = \hat{f}(\omega)\hat{h}(\omega)$.
3. Perform inverse transform to obtain the deconvoluted signal: $\hat{g}(\omega) \Rightarrow g(t)$.

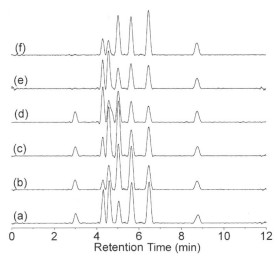

Figure 5.59. Baseline-corrected WPT coefficients after subtracting the estimated baseline by linking the minimum point at both sides of the peak in Figure 5.58.

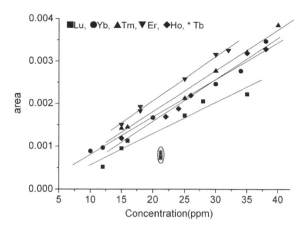

Figure 5.60. Calibration curves obtained from the peak area in Figure 5.59 and concentrations of each component in the samples.

Generally, for discrete signals, FFT is used for calculation, and the window functions include the box window, the triangle window, the Hanning window, and the Hamming window. In practice, the width and position of the window must be determined by trial and error, which is always time-consuming.

Figure 5.61 compares the results obtained by two FFT deconvolution methods and the WT method of the simulated and experimental NMR spectra in curves (a) of Figures 5.52 and 5.56, respectively. Curves (a1) and (b1) in Figure 5.61 are obtained by the FFT method on the basis of the abovementioned procedures, where the Hanning window is used, those coefficients out of the window are cut off. Curves (a2) and (b2) are obtained by an improved FFT method based on the abovementioned procedures, where all the coefficients remain but those coefficients in the Hanning window are amplified by a factor k. It should be noted that many trials must be done in order to obtain satisfactory results by FFT methods, because we must select a suitable window width and position. For the second method, we must determine a suitable value for the parameter k. The (a3) and (b3) curves are obtained by the WT method in exactly the same way as curves (b) in Figures 5.52 and 5.56. In this method, only the value of the parameter k should be optimized by trial and error.

From the results of the simulated spectrum, it can be seen that we cannot obtain a satisfactory baseline by the first FFT method, because the low-frequency part of the signal is cut off. For the second FFT method and the WT method, the baseline will not change because all the information remain, and the resolution enhancement is caused by the increase of the

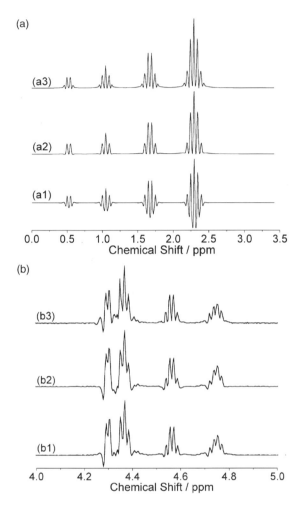

Figure 5.61. Comparison of the results from the simulated (a) and experimental (b) NMR spectra by FFT deconvolution (a1, b1, a2 and b2) and WT method (a3 and b3).

high-frequency part of the signal. We can easily control the resolution of the results by changing the value of the parameter k. Comparing curves (a2) and (a3) in Figure 5.61a, the improved FFT method is even superior to the WT method because there are positive sidelobes in (a3). From the results of the experimental spectrum, it can be seen that there is no significant difference among the results of the three methods, and the resolution of (b3) is slightly better than that of curves (b1) and (b2) in Figure 5.61. Furthermore, it should be noted that we cannot expect to further enhance the resolution by using a higher value of k in the second FFT method because the negative sidelobes in (b2) will increase with the value of the parameter k.

5.5. COMBINED TECHNIQUES

By the applications of WT discussed above, it has been shown that WT is a powerful tool for compression, denoising, and resolution of analytical signals. In the following text, we will discuss the extended applications based on the combined techniques of WT and other chemometrics methods.

5.5.1. Combined Method for Regression and Calibration

Regression and calibration are the most commonly used techniques for quantitative determination in analytical chemistry. There is no difficulty in regression and calibration calculations for traditional analytical methods because only one variable is used. In modern analytical chemistry, spectroscopic methods are increasingly employed for quantitative determination. New methods such as multiple linear regression (MLR) for modeling multivariable datasets are needed. However, the dimensionality of spectral datasets is basically limited by the number of the objects studied, whereas the number of variables can easily reach a very large number. Furthermore, the high-dimensional spectral data are closely correlated and usually noisy. Therefore, methods more suitable for modeling correlated variables are proposed, such as the principal-component regression (PCR) and partial least-squares (PLS) methods.

Generally, all the information contained in the spectra can be used for the modeling; these are called *full-spectrum methods.* However, in many cases, a preprocessing of the experimental spectra using WT compression can offer some advantages compared to the full-spectrum methods.

A combined procedures of WT compression and PLS is illustrated in Figure 5.62, including the following steps:

1. The measured signals, denoted by **X**, such as spectra, are transformed into wavelet domain represented by wavelet coefficients, **W**.
2. The matrix **W** is sorted according to their contribution to the data variance and a matrix $\mathbf{W}_{sorted}$ can be obtained. Because many wavelet coefficients in **W** or $\mathbf{W}_{sorted}$ are usually very small, only a limited number of columns of $\mathbf{W}_{sorted}$ are needed to represent the signal **X**. Therefore, the $\mathbf{W}_{sorted}$ can be divided into two submatrices, $\mathbf{W}_s$ and $\mathbf{W}_n$, containing significant (information component) and insignificant (noisy component) coefficients, respectively. This step can be skipped in many cases because the sorting will change the relative position of the coefficients and, subsequently, cause a variation of the original information.

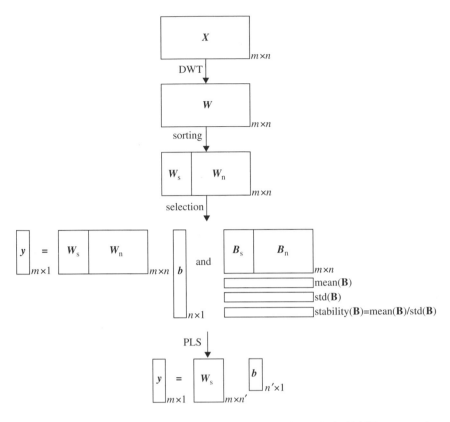

Figure 5.62. A diagram showing the procedures of the PLS coupled with WT compression.

3. The submatrix $\mathbf{W}_s$ can be determined by different criteria:

   3a. We may simply use the criteria discussed in the WT compression, but this means that only the advantage of WT compression is utilized.

   3b. Other methods can also be employed for this purpose, such as the relevant component extraction (RCE) PLS approach described in Walczak's book, *Wavelets in Chemistry* [12]. In this method, as illustrated in Figure 5.62, the PLS is employed to calculate the **b** coefficients. A matrix of the regression coefficients can be obtained by using the "leave one out" cross-validation procedure. Then, the stability of the regression coefficient i, defined by

$$\text{Stability}(b_i) = \frac{\text{mean}(b_i)}{\text{std}(b_i)} \qquad (5.81)$$

can be calculated. Using the maximal stability of the noisy variables as a threshold,

$$\text{Threshold} = \max\left(\text{abs}(\text{stability}_{noise})\right) \qquad (5.82)$$

we can cut off those coefficients in **b** and the corresponding wavelet coefficients in the **W** or **W**$_{sorted}$.

4. With the submatrix **W**$_s$, build the PLS model from the training set.
5. Finally, we can use the model for prediction. It should be noted that the experimental data must be processed in the same way as that of the training set used to build the model. When you use step 3a for compression, you must compress the experimental data with the same criteria. When you use step 3b for determination of the **W**$_s$, you should keep those coefficients at the same position.

This method has been successfully used in the analysis of NIR spectra of gasoline samples. Examples can be found in the book *Wavelets in Chemistry* [12].

5.5.2. Combined Method for Classification and Pattern Recognition

Generally, *pattern recognition* refers to the ability to assign an object to one of several possible categories according to the values of some measured parameters, and the classification is one of the principal goals of pattern recognition. Many methods have been proposed for classification and pattern recognition because of their importance in chemical studies. Combined methods of WT for classification and pattern recognition include two main steps: (1) compression or feature selection is performed to the original dataset using WT as a preprocessing technique; then (2) classification or pattern recognition is performed by classifiers such as the artificial neural network (ANN), the soft independent modeling of class analogy (SIMCA), and the *k*th nearest neighbors (KNN), in the wavelet domain.

There have been several successful examples based on the combined method of WT and conventional classifiers for classification of analytical signals. One of them is reported by Bos and Vrielink in *Chemometrics and Intelligent Laboratory Systems*, [**23**:115–122, (1994)]. In their report, identification of mono- and disubstituted benzenes utilizing WT and several classifiers from IR spectra was studied. The aim of their work is to show whether the localization property of WT in both position and scale can be used to extract this information into a concentrated form to obtain the salient features of an IR spectrum effectively. The coefficients obtained from the WT treatment of the IR spectrum were employed as inputs for an identification process that is based on the linear or nonlinear neural network classifiers. Using the concentrated form instead of the full spectra, the time to develop the classifiers is greatly reduced. Moreover, it is expected that the quality of the classifiers will improve if they are derived from smaller datasets that contain all the relevant information. From their study, it is concluded that WT coupled with Daubechies wavelet functions is a feature extracting method that can successfully reduce IR spectral data by more than 20-fold with a significant improvement in the classification process.

Another example is reported by Collantes et al. in *Analytical Chemistry*, [**69**:1392–1397 (1997)]. They employed WPT for preprocessing of HPLC data and several classifiers as potential tools for pharmaceutical fingerprinting pattern recognition. The HPLC data for each L-tryptophan sample was preprocessed by the Haar wavelet function in the WPT treatment. Then, the coefficients thus obtained in the wavelet domain were sorted in descending order. A small portion of sorted coefficients were fed as the inputs of the ANN, KNN, and SIMCA classifiers to classify the samples according to manufacturers. With this study, they concluded that WPT preprocessing provides a fast and efficient way of encoding the chromatographic data into a highly reduced set of numerical inputs for the classification models.

For purposes classification and regression, an adaptive wavelet algorithm (AWA) using higher-multiplicity wavelets was proposed. Detailed descriptions of the method can be found in Chapter 8 and Chapter 18 of the book *Wavelets in Chemistry* [12]. The wavelet neural network (WNN) can also be used for classification and pattern recognition, discussed in the following section.

5.5.3. Combined Method of Wavelet Transform and Chemical Factor Analysis

There are several ways to combine WT and chemical factor analysis (CFA) for different purposes. In Section 5.3.5, for example, WT is used for background removal in order to obtain the correct rankmap of a data matrix.

Principal-component analysis (PCA) is an important and basic method in CFA. However, because the principal components (PCs) are calculated as the eigenvectors of the variance–covariance matrix, computation of the

PCs is time-consuming. In order to speed up the calculation of PCA, a fast approximate PCA has been proposed using WT or WPT as a tool for data compression. Generally, as can the procedures in the combined methods for classification and pattern recognition, the main procedures of the fast approximate PCA can be outlined as two steps: to compress the data set and then perform the PCA. As an example, one of the algorithms for WPT compression and PCA of the set of spectral signals can be summarized as follows:

1. Calculate the "variance spectrum" by

$$\text{var}_j = \frac{1}{m} \sum_{i=1}^{m} (x_{ij} - \bar{x}_j)^2 \qquad (j = 1, \ldots, n) \qquad (5.83)$$

where m denotes the number of objects, n denotes the number of variables, x_{ij} denotes an element the dataset $\mathbf{X}_{m \times n}$, and $\bar{x}_j$ is the mean of the jth column calculated as

$$\bar{x}_j = \frac{1}{m} \sum_{i=1}^{m} x_{ij} \qquad (5.84)$$

2. Decompose the variance spectrum into the WPT coefficients represented by the WPT tree.
3. Search the WPT tree for the best basis.
4. Compress the coefficients according to a selected criterion.
5. Decompose all the spectra into WPT coefficients and compress them in the same way as in steps 2 and 4.
6. Perform PCA on the compressed coefficients.

Using WT or WPT as a denoising tool for CFA is another type of combination of WT and FA. An example for resolution of multicomponent chromatograms by window factor analysis (WFA) with WT preprocessing can be found in the *Journal of Chemometrics* [**12**:85–93 (1998)]. In this paper, resolution and quantitative determination of a multicomponent chromatogram with strong noise was investigated. It was proved that both the resolved chromatographic profiles and the quantitative results by WFA can be greatly improved for the noisy chromatographic data matrix by WT preprocessing. Figure 5.63a,b shows the resolved chromatograms both without and with the WT preprocessing, respectively. The dotted (elliptical) lines in the figure are the experimental chromatograms

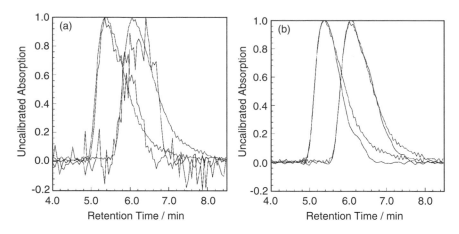

Figure 5.63. Resolved chromatograms by WFA without (a) and with (b) the WT preprocessing. The dotted lines represent the experimental chromatograms of the standard samples for comparison [from *J. Chemometr.* **12**:85–93 (1998)].

of the standard samples. From this figure we can see the effect of WT preprocessing.

5.5.4. Wavelet Neural Network

The wavelet neural network (WNN) is a combination of wavelet transform and the artificial neural network (ANN). WNN uses wavelet functions instead of the traditional sigmoid function as its transfer function in each neuron. Two different models have been proposed for different applications: (1) one is for general purposes such as quantitative prediction, classification, and pattern recognition and (2) one is for signal compression. Their architectures are illustrated in Figure 5.64a,b, respectively. In the first model, the architecture is almost exactly the same as ANN except that the transfer function is replaced by a wavelet function $\psi_{a,b}(t)$. In the second model, the input is a parameter t_i describing the position of the compressing signal, such as the wavenumber for IR or NIR and retention time for the chromatogram. In some of the literature, this model is regarded as one input neuron. It would be more accurate to say that there is no input layer in the model because the input parameter t_i is directly fed into the middle layers' neurons without any processing. Furthermore, there is only one output neuron in the model because the output is the magnitude of the signal at the position t_i.

For both models, the learning procedures are similar with the traditional ANN, namely, to modify the parameters w_{ik}, w_{kj}, a_k, and b_k for model (a)

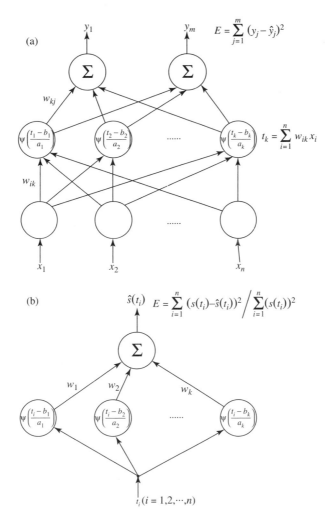

Figure 5.64. The architecture of the WNN for general purposes (a) and signal compression (b).

(in Fig. 5.64) and w_k, a_k, and b_k for model (b) according to the value of the output error by the conjugate gradient method.

The most commonly used wavelet is the Morlet wavelet basis function in the WNN, which is defined as

$$\psi(t) = \cos(1.75t) \exp\frac{-t^2}{2} \tag{5.85}$$

Applications of the two WNN models have been studied, such as pattern recognition of overlapping UV–vis spectra, classification of analytical

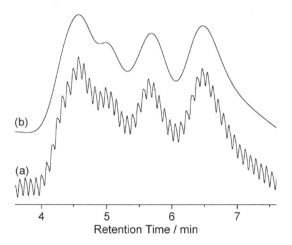

Figure 5.65. An experimental chromatogram (a) and the reconstructed results from the WNN parameters (b).

samples based on analytical data, quantitative determination from over-lapping UV–vis spectra, and compression of the IR spectrum, which can be found in the literature.

Figure 5.65 shows an example using model (b) (of Fig 5.64) for compression of a chromatogram. Curve (a) in Figure 5.65 shows a chromatogram composed of 800 sampling points, and curve (b) was obtained by reconstruction of 12 wavelet bases optimized by WNN. Thus, we can represent the chromatogram only by 36 parameters. The compression ratio is more than 20 : 1. In this example, there is a high level of noise in the original experimental chromatogram. However, in the reconstructed chromatogram, the noise is filtered out. This can be explained by the fact that a limited number of wavelet bases is used in the reconstruction, and it represents only the primary information of the chromatogram. Much more wavelet basis will be needed to represent the high-frequency noise. Therefore, if we control the parameters of the WNN properly, the method can be used for compression and denoising simultaneously.

5.6. AN OVERVIEW OF THE APPLICATIONS IN CHEMISTRY

As mentioned at the beginning of this chapter, more than 370 papers have been published on WT. Applications of WT in chemistry have been extensively studied. According to the statistics of the published papers on this topic by Dr. Alexander Kai-man Leung (please refer to his homepage at *http://fg702-6.abct.polyu.edu.hk/wavelet.html*), WT has been applied in the following areas:

- Analytical chemistry

  Capillary electrophoresis (CE)
  Chemiluminescence/fluorescence spectroscopy
  Flow injection analysis (FIA)
  High-performance liquid chromatography (HPLC)
  Gas chromatography (GC)
  Graphite furnace atomic absorption spectrometry (GFAA)
  Inductively coupled plasma atomic emission spectrometry (ICP-AES)
  Infrared spectrometry (IR)
  Mass spectrometry (MS)
  Nuclear magnetic resonance (NMR) spectrometry
  Potentiometric titration
  Photoacoustic spectroscopy (PAS)
  Raman spectroscopy
  Ultraviolet–visible spectrometry (UV–vis)
  Voltammetry
  X-ray diffraction/spectroscopy

- Chemical engineering
- Chemical physics
- Quantum chemistry
- Miscellaneous (denoising, compression, comparison studies and review papers, etc.)

In the following sections, the published papers are briefly reviewed to give the readers an overview of the applications of the WT techniques in different fields.

5.6.1. Flow Injection Analysis

As early as in 1992, Bos and Hoogendam [13] proposed using WT to minimize the effect of noise and baseline drift in flow injection analysis. When a FIA system is operated near the detection limits, it is difficult to locate peaks and find the right baseline correction method, because weak signals are embedded in the stochastic noise. In this work, the FIA signal was transformed into a two-dimensional time–frequency form so as to obtain the denoised peak intensity optimally. The maximum peak position can be obtained by searching the wavelet coefficients of the FIA signal obtained for a sample with a relatively high concentration. By using the peak intensity at the maximum peak position, quantitative determination can be further performed. Both the simulated peaks of Gaussian and exponentially modified Gaussian (EMG) and the experimental signals of Imidazole samples and 18-crown-6 samples were studied. Results show that, for a white noise and a favorable peak shape, a signal-to-noise ratio (SNR) of 2 can be tolerated at the 5% error level. This means that a significant reduction in the detection limit can be obtained in comparison with the conventional signal processing methods.

The character of baseline noise strongly influences several steps in the data processing procedure, including choice of the optimal filter, peak detection, peak boundary setting, precision of integration results, peak purity detection, and choice of the calibration method. Generally, noises can be classified into groups such as white noise, heteroscedastic noise, and correlated noise ($1/f$ noise), which have different properties and subsequently have different effects on the processing of analytical signals. The second paper published in FIA is reported by Mittermayr et al. [14]. The goal of the paper is to detect correlated noise from the actual measurement in the presence of peaks without the necessity of the additional measurements of pure baselines and assumptions about the ergodicity. On the basis of the ability of WT to provide information in both time and frequency domains, the authors used the wavelet power spectral density (WPSD) to get a low-resolution equivalent to the traditional power spectral density (PSD) based on the Fourier transform (FT). Then the work demonstrated that the WT separates signal and noise in both the time and frequency domains simultaneously. An F-test is proposed to detect the presence of correlated noise, and the correlation parameter is estimated by weighted least square regression. Finally, the WSPD is applied to flow injection analysis (FIA) signals and its results are compared to the estimates obtained from data with baseline noise.

5.6.2. Chromatography and Capillary Electrophoresis

More than 30 papers have been published on adopting WT in chromatographic data processing, including smoothing, denoising [15–20], data compression, [19,20], baseline correction [21,22], resolution of overlapping chromatograms, [23–27], and classification studies with the combined techniques [28].

Smoothing and denoising is a universal problem in signal processing. In chromatographic data processing, noise suppression is also a very common technique. It attempts to improve a chromatogram to give a higher signal-to-noise ratio (SNR). Traditionally, techniques such as the Fourier and Kalman filters and the Savitsky–Golay method are generally used for smoothing and denoising. As discussed in Section 5.1.2, WT is a very good technique for this purpose because of its ability to decompose a signal into components of different frequency. Applications similar to those described in Examples 5.4–5.7 can be found in the literature [15–17].

The work reported by Mittermayr et al. [18] used WT as a tool to improve the calibration and the detection limit of a gas chromatograph coupled with a microwave-induced plasma detector system. Symmlet wavelets and universal soft thresholding were employed for denoising. By comparing the technique with the Fourier and the Savitsky–Golay filters, they concluded that special care should be taken for the choice of the filter parameters, especially when large variations in peak widths are present. In such cases, denoising with wavelet filter was better than with the other two methods.

In References [19] and [20], two combined techniques for denoising and compression of chromatograms were proposed. The first one is the WNN introduced in Section 5.5.4. The second one is a combined technique of WT and a genetic algorithm (GA). The underlying theme of both papers was to represent an analytical signal by linear combination of wavelet functions

$$\hat{f} = \sum_{i=1}^{N} c_i \psi_{a_i,b_i} \tag{5.86}$$

where N is the number of elementary functions, ψ_{a_i,b_i} is a set of elementary functions defined by dilation with a_i and translation with b_i, c_i is the corresponding coefficient, and $\hat{f}$ can be regarded as an approximation of the original signal f. The genetic algorithm is applied to find all the parameters a_i, b_i, and c_i that fit the original signal best. B2-spline, B3-spline, Marr, and Morlet basis functions were studied. The B2-spline was found to be in computation and precise in representation. The number of the elementary function, N, is an important parameter in the method because it determines the compression ratio and the effect of denoising. However, the parameter depends on the complexity of the signals being analyzed and must be determined by manual trials.

References [21] and [22] and Sections 5.3.3 and 5.3.5 (in this chapter) discuss methods for removal of baseline or background from 1D and 2D chromatograms. WT-based techniques for resolution enhancement of overlapping chromatograms are discussed in Sections 5.4.4 and 5.4.5. Both WT and WPT can be used for this purpose. These techniques are described in further detail in References [23–25]. In Reference [26], the technique is applied to plant hormone analysis. Plant hormones eluted as overlapping chromatographic peaks were quantitatively determined with the help of WT resolution. Application of CWT for resolution enhancement was also reported [27]. Compared with the DWT, CWT is a more redundant transform. It tends to reinforce the traits of a signal and makes all information more visible because of its redundancy. Therefore, CWT has much more capable of extracting subtle information from seriously overlapping signals. Furthermore, in CWT, the parameters a and b vary continuously; we can choose the exact value of a to depict the component of a certain frequency band in which we are interested, whereas in DWT, we cannot adjust the analytical window precisely to meet a particular need. In this study, the magnitude of each coefficient is represented by graytone and the coefficient matrix was plotted in a graytone graph. Darker spots correspond to larger coefficients. The highest-resolution chromatogram can be obtained by visual inspection of the graytone graph.

Application of WPT combined with classifiers was also studied by Collantes et al. [28] In their work, several computer-based classifiers were evaluated as potential tools for pharmaceutical fingerprinting, which was based on analysis of HPLC trace organic impurity patterns using WPT compression.

An online WT technique for denoising and resolution of overlapping chromatograms was proposed in References [29] and [30]. At first, a "parallel algorithm" calculating $\mathbf{c}_j$ and $\mathbf{d}_j$ for $j = 1, \dots, J$ simultaneously with the progress of sampling was proposed, which can be described as follows:

```
/* Prepare hⱼ and gⱼ */
/* Preset the best resolution level J */
While (!stop-sampling)
   /* Wait until elapsed time equals to the sampling
      time interval */
   /* Sample the kth point of cⱼ */
   for ( j=1; j<=J; j++)
```

$$c_{j,k} = \frac{1}{\sqrt{2}} \sum_{m=0}^{L-1} h_{j,m} c_{j-1,k-m} \quad \text{/* } L \text{ is the length of the filters */}$$

$$d_{j,k} = \frac{1}{\sqrt{2}} \sum_{m=0}^{L-1} g_{j,m} c_{j-1,k-m} / * c_{j-1,k-m} = 0 \text{ when } k - m < 1 * /$$

```
    /* Graphical display of d_{J,k} or c_{J,k} and d_{j,k} */
  end;
    k = k + 1;
End of while;
```

where k runs from 1 to N ($k = 1, \ldots, N$) and N is the total number of the sampling point.

Using this program, online denoising and resolution enhancement of high-performance liquid chromatography (HPLC) were studied. In these studies, the signal from detector was sampled by an A/D (analog-to-digital) convertor, and $c_{j,k}$ and $d_{j,k}$ can be obtained by the program simultaneously with the progress of sampling. With mixed samples using benzene, methyl benzene, and ethyl benzene, performance of the method in denoising and resolution enhancement was investigated in the References [29] and [30], respectively. Eight samples with low concentration were measured in the denoising study. The correlation coefficient above 0.99 was obtained for five standard samples, and the recoveries of three mixed samples lay between 94.0 and 105.0%. Comparison of the results with those obtained from the unprocessed chromatograms indicated that the quality of the chromatographic signals was improved by the online wavelet transform. In the study of resolution enhancement, quantitative calculation of three samples was also investigated, and recoveries between 96.3 and 104.5% were obtained.

Applications of WT in capillary electrophoresis analysis are similar with those methods used in chromatography studies, including smoothing, denoising, and resolution of the overlapping peaks [31–33]. Schirm et al. [33] developed a method to transfer the use of fingerprints from spectroscopy to electrophoresis by interpolation and wavelet filtering of the baseline signal. The resulting data were classified by several algorithms, including self-organizing maps (SOMs), artificial neural networks (ANNs), soft independent modeling of class analogy (SIMCA), and k nearest neighbors (KNNs). In order to test the performance of this combined approach in practice, it was applied to process the data from the quality assurance samples of pentosan polysulfate (PPS). A capillary electrophoresis method using indirect UV detection was employed. All algorithms were capable of classifying the examined PPS test batches. Even minor variations in the PPS composition not perceptible by visual inspection could be automatically detected. The entire method has been validated by classifying various unknown PPS quality assurance samples, which have been correctly identified without exception.

5.6.3. Spectroscopy

The spectroscopic techniques include atomic absorption spectroscopy (AAS), atomic emission spectroscopy (AES), fluorescence spectroscopy, ultraviolet–visible (UV–vis) spectroscopy, infrared (IR) spectroscopy, near-infrared (NIR) spectroscopy, Raman spectroscopy, nuclear magnetic resonance (NMR) spectroscopy, X-ray spectroscopy, and photoacoustic spectroscopy (PAS). These techniques have been widely used in analytical chemistry for both qualitative and quantitative analysis. As in other analytical techniques, the raw spectral data of spectroscopic measurements are a combination of useful signals from the analyzing sample and noise from various forms of interference. Signal processing methods are commonly used to extract the useful signal from the raw experimental data.

Applications of WT-related methods have been studied extensively in spectroscopic signal processing. About 80 papers have been published on these topics. Only some of the papers are reviewed here.

Infrared spectroscopy plays an important role in the identification and characterization of chemicals. Applications of WT in IR spectroscopy can be categorized mainly into denoising [34], compression [35–37], pattern recognition [38], signal extraction [39], and variable selection for PLS regression [40].

Alsberg et al. [34] reported a comparative study in applying WT to denoise IR spectra. Six different methods, including SURE, VISU, HYBRID, MINMAX, MAD, and WPT, were applied to pure IR spectra with the addition of different levels of homo- and heteroscedastic noise. Results show that at higher SNR, the wavelet denoising methods were better, especially, the HYBRID and VISU methods. However, at very low SNR, there was no significant difference between the performance of the wavelet methods and the traditional methods, such as Fourier and moving mean filtering. This study concluded that the visual quality of WT denoised infrared spectra is superior.

Several papers were published on the compression of IR spectrum utilizing WT. Reference [35] proposed a successful method for the compression of experimental infrared spectra. This method was based on FWT coupled with multiresolution signal decomposition (MRSD) as well as the optimal bit allocation quantization and Huffman coding techniques. Comparison of the results of IR spectra using several chemicals from the method proposed in Reference [35] with results from FFT and wavelet-based threshoding methods revealed that the proposed method outperforms the other two.

Leung et al. [36] used WT to reduce IR spectral data for library searching. In this work, FWT and WPT were applied to compress infrared IR spectrum for storage and spectral searching. The coefficient position-retaining (CPR) method was discussed in detail for handling data with any length. IR spectra of 20 organic compounds with similar structures were compressed at the fourth resolution level with the use of the Daubechies wavelet function ($L = 16$). After data compression, the coefficients obtained were selected and employed to build a spectral library for future searching. Spectral library searching of this database was found to be better than that treated by FFT, especially in the aspect of visual comparison in some cases. The scale coefficients obtained from FWT and WPT can be used effectively for preliminary searching in a large spectral library. The proposed methods can minimize the search time by using a direct matching method.

Application of WNN to compression of IR spectrum was also reported [37]. In this study, the Morlet wavelet function was employed as the transfer function. The wavenumber and the transmittance of the IR were chosen as the input and output of the network, respectively. With proper training, the weighting factor for each neuron and the parameters of the wavelet function can be optimized. The original spectrum can be represented and compressed by using the optimized weighting factor and wavelet function parameters. In this work, compression ratios of 50 and 80% were obtained when the wavenumber interval of the IR spectrum was 2.0 and 0.1 cm^{-1}, respectively.

Bos and Vrielink [38] reported their results on identification of mono- and disubstituded benzenes utilizing WT preprocessing. The aim of the work was to examine whether the localization property of WT in both position and scale can be used to extract features of an IR spectrum effectively. After the WT treatment on the IR spectrum, the coefficients obtained were employed as inputs for an identification process based on linear and nonlinear neural network classifiers. It was shown that, by using the extracted information instead of the full spectrum, the classification was greatly improved in both speed and quality. This study was also showed that WT with the Daubechies wavelet is a good method for feature extraction that can reduce IR spectral data by more than 20-fold.

The Fourier transform IR spectrum of a rock contains information about its constituent minerals. Stark et al. [39] employed WT to predict the mass fraction of a given constituent in a mixture. Using the wavelet transform, the authors roughly separated the mineralogical information in the FT-IR spectrum from the noise, using an extensive set of known mineral spectra as the training data for which the true mineralogy is known. Wavelet coefficients that varied either too much or too little were ignored because the former coefficients are likely to reflect analytical noise and the latter coefficients do not help one discriminate between different minerals. The remaining coefficients were used as the data for estimating the mineralogy of the sample. In this work, an empirical affine estimator was also developed to estimate the mass fraction of a given mineral in a mixture. The estimator was found to typically perform better than the weighted nonnegative least-squares instrument.

Variable selection and compression are often used to produce more parsimonious regression models. When they are applied directly to the original spectrum domain, however, it is not easy to determine the type of feature that the selected variables represent. In another study, Alsberg et al. [40] showed that it is possible to identify important variables as being part of short- or large-scale features by performing variable selection in the wavelet domain. The suggested method can be used to extract information about the selected variables that otherwise would have been inaccessible, and to obtain information about the location of these features in the original domain. In this article, three types of variable selection methods were applied to the wavelet domain: selection of optimal combination of scales, thresholding based on mutual information, and truncation of weight vectors in the PLS regression algorithm. It was found that truncation of weight vectors in PLS was the most effective method for selecting variables. Two experimental datasets were investigated. Results showed that approximately the same prediction error was obtained by using less than 1% and 10% of the original variables, respectively. In this work, it was also found that the selected variables were restricted to a limited number of wavelet scales. This information can be used to suggest whether the underlying features may be dominated by narrow peaks (indicated by variables in short-wavelet-scale regions) or by broader regions (indicated by variables in long-wavelet-scale regions). This study also concluded that the variables selected are not unique when the variable selection is applied to collinear data such as spectral profiles of complex mixtures. In most cases, we cannot expect to find a very limited number of unique variables, but rather regions of interest where good representative wavenumber candidates are found. This suggests that instead of performing the variable selection in the original domain, a compressed domain representation may be more fruitful.

Near-infrared (NIR) spectroscopy has become increasingly popular in quantitative determination. Variable selection and compression are often used to build better regression models. Trygg and Wold [41] studied PLS regression on wavelet compressed NIR spectra and showed that, using DWT as a preprocessing method in regression modeling, compression of the data up to 3% of the original size can be achieved without any loss of information. The predictive ability of the compressed regression model is basically the same as that of the original uncompressed regression model. Furthermore, a method for quantitative analysis of NIR spectra by wavelet coefficient regression with the help of a genetic algorithm [42] and application of linear regression on wavelet coefficients for robust calibration of spectral data with a highly variable background [43] were proposed, and the results are encouraging. In the latter study [43], the Monte Carlo technique was used to investigate the performance of the method in cases where the background variation in the prediction set was both (1) the same as and (2) differed from that in the calibration set. Multivariate linear regression on wavelet coefficients proved to be competitive in the case 1 and superior in case 2 with respect to the partial least squares (PLS) calibration. This work is a good example of how the properties of WT can be used for reducing the effects of varying background. As a background correction method, the proposed approach avoided errors introduced in the estimation process.

In analytical chemistry, especially in spectroscopic studies, data transfer (i.e., the comparison of the performance between analytical instruments) or calibration transfer (i.e., where a calibration model developed on one instrument is transported to other instruments) is a very important problem. Several methods have been proposed for the purpose, such as slope and bias correction (SBC), direct standardization (DS), piecewise direct standardization (PDS), orthogonal signal correction (OSC), finite impulse response (FIR) filtering application of neural networks, and WT-based techniques. Walczak et al. [44] proposed a new standardization method for comparing the performance between two NIR spectrometers in the wavelet domain. They tried to relate the WT coefficients of the NIR spectra obtained from two different instruments utilizing an univariate linear model. WT was applied on the NIR spectra obtained from both the "master" (first) spectrometer and the "slave" (second) spectrometer. Then, an univariate linear model was set up to determine the standardization parameters between the two sets of NIR spectra. Once these parameters were found, NIR spectra in the wavelet domain could be transferred between two different spectrometers for subsequent data analysis.

Another standardization method called *wavelet hybrid direct standardization* (WHDS) was proposed by Tan and Brown [45]. It is based on a new wavelet reconstruction algorithm in which approximation and detail spectra are reconstructed separately. Piecewise direct standardization (PDS) and direct standardization (DS) were used to correct the differences of original spectra by transforming the reconstructed approximation and detail spectra, respectively. The proposed method is applied to NIR data, and its performance is compared with that using conventional methods. Results showed that, using the wavelet multiresolution technique and combining PDS and DS, the WHDS algorithm allows a more robust and reliable means for standardization when transfer standards are available. More practical examples can be found in References [46] and [47].

An automated method integrating wavelet processing and techniques from multivariate statistical process control (MSPC) is reported by Stork et al. [48]. This method provided a means of simultaneous localization, detection, and identification of disturbances in spectral data. Because of the ability of WT to map a one-dimensional spectrum into a two-dimensional function of wavelength and scale, the method employing WT processing results in the generation of multiple models within the wavelength-scale domain. Provided the spectral disturbance can be localized within a subregion of the wavelength-scale domain through an advantageous choice of basis set, the method allows identification of the underlying disturbance.

Denoising and spike removal of Raman spectra were reported by Cai [49] and Ehrentreich and Summchen [50], respectively. In the former work, the Raman spectra of the alcohol solution of CCl_4 with different concentrations are measured and denoised by WT. The work showed that WT is very efficient in extracting weak signals from high-level noise background and maintains the linear relationship between the peak intensity and the concentration of the solutions. The results indicated that WT can be used in quantitative analysis of the weak signal. In the latter work [50], a method for spike removal was proposed. Although the suppression of spikes is not straightforward by WT, WT can be used to recognize the spikes by their first-level detail coefficients. Then, spike locations can be projected from the details to the approximations and, further, to appropriate locations of the original spectrum. After recognition, the spikes can be removed by replacing those regions by interpolated values.

Nuclear magnetic resonance (NMR) spectroscopy is one of the most powerful techniques for probing structures of chemical compounds. A few papers have been published on the application of WT in NMR spectroscopy. Neue [51] published a paper on application of WT in dynamic NMR spectroscopy that could simplify the analysis of the free induction decay (FID) signal. The localization property of WT gives a better picture of the nature of the underlying dynamical process in both the frequency and time domains.

CWT was also studied as an analysis tool in NMR spectroscopy by Barache et al. [50], who adopted CWT for removal of a large spectral line and rephasing the NMR signal influenced by eddy currents. The theory and examples have been discussed in Section 5.3.4.

Serrai et al. [53] published another paper on the application of WT in time-domain quantification of NMR parameters, including amplitude, chemical shift, apparent relaxation time T_2^*, and phase. The proposed method separates each component from the FID first by using WT, then successively quantifies it and subtracts it from the raw signal. Both simulated and experimental data were tested by the method; results indicated that the WT method can provide efficient and accurate quantification of NMR data. Compression and resolution of NMR signals were also reported [54,55] and are discussed in Sections 5.1 and 5.4.

Because ultraviolet–visible (UV–vis) spectroscopy is an important tool for characterization, identification, and quantification of substances, applications of WT in UV–vis spectroscopy have been extensively studied. Denoising and compression of UV–vis spectra were studied in several papers [56–58], followed by pattern recognition [59] and resolution of overlapping peaks [60].

WNN was applied by Liu et al. [59] to recognize the UV–vis spectra of tyrosine, 3,4-dihydroxyphenylalanine, and trytophane. The Morlet wavelet and line search conjugate gradient optimization method were used in their neural network. The results indicated that the wavelet neural network had a very good recognition power to differentiate minor differences between similar UV–vis spectra.

Ren and Gao [60] adopted WT for removing noise and irrelevant information from spectrophotometric spectra. A PLS based on the wavelet multiresolution analysis (WPLS) method was developed to perform simultaneous spectrophotometric determination of Fe(II) and Fe(III) with overlapping peaks. Results showed the WPLS method to be successful even for spectra with severe overlapping. Data reduction was also performed using wavelet multiresolution analysis and the principal-component analysis (PCA) algorithm. Results indicated that this method can be considered as a powerful tool for efficient compression of experimental data and can be applied for rapid simultaneous multicomponent determination.

Photoacoustic (PA) spectroscopy is a comparatively new technique in analytical chemistry. It is a complementary technique of conventional spectroscopy for studying those materials that are unsuitable for the transmission or reflection methodologies. The application of WT in PA spectroscopy concentrated mainly on noise filtering, baseline separation, and resolution enhancement [61–63]. Mao et al. [61] applied WT to analyze the PA spectra of degraded poly(vinyl chloride) (PVC). Shao et al. [62] reported a technique for denoising and resolution of PA spectra using the online WT algorithm discussed in Section 5.6.2. In Reference [63], a WT-based derivative calculation was used for enhancement of the PA spectra.

Applications of WT in X-ray spectroscopy [64–66], atomic absorption spectroscopy (AAS) [67], atomic emission spectroscopy (AES) [68], fluorescence spectroscopy [69] were also reported. Detailed information can be found in the reference cited above.

5.6.4. Electrochemistry

Electrochemistry is an important branch of chemistry that plays a very important role in many areas such as chemical analysis, thermodynamic and kinetic studies, electrochemical synthesis, and energy conversion, as well as biological electron transport. Electroanalytical techniques include conductivity, potentiometry, voltammetry, amperometric detection, and coulometry. Electrochemistry is another field where WT was extensively studied. More than 50 papers have been published on the subject. These applications can be divided into smoothing and denoising, useful information retrieval, resolution of overlapping signals, and quantitative determination using combined techniques. Since these techniques have been discussed previously, only several selected papers are reviewed here.

In applications of WT in electrochemistry studies, the spline wavelet is the major wavelet function, which is rarely utilized in other fields. Details of the spline wavelet can be found in Chapter 4 and in References [4–12].

As in many fields, smoothing and denoising represented the first application of WT in processing the electroanalytical signals. In the paper by Yan and Mo [70], a real-time continuous wavelet was developed to denoise signals from the staircase voltammetry. Fang and Chen [71] proposed a new tool for processing electroanalytical signals with an adaptive wavelet filter based on the WPT technique. Their study showed that the wavelet adaptive filter can be applied to a system in which the interference originated from the power supply. This is useful for the study of fast-electron-transfer processes. Fang et al. [72] also proposed a new algorithm for decomposition calculation without the limit of data point number, and the method was successfully used in extracting weak signals. Applications of WT in recovering useful information from different kinds of oscillographic chronopotentiometric signal were also studied [73,74].

Another combination technique was developed by Zheng et al. [75,76], who coupled the spline wavelet and the Riemann–Liouville transform (RLT) together to filter random noise and extraneous currents in voltammetric signals. They processed both simulated and experimental data. The results showed that signals of SNR $= 0.8$ can be filtered. The errors of the peak current were less than 5% and those of the peak potential were less than 1%.

WT has been proposed as an alternative tool to overcome the limitations of FFT in the analysis of electrochemical noise measurements (ENM) data by Aballe et al. [77], who applied both WT and FFT methods to study various different corrosion systems covering a wide range of ENM signals. Results demonstrated that WT is applicable to those systems in which the

FFT technique works. However, in cases where FFT fails, WT can still provide valuable knowledge about the behavior of the system. Using WT, the different ENM components contributing to the original signal can be characterized. Each component is defined by a set of wavelet coefficients, which contain information about the timescale characteristic of the associated corrosion event.

In differential pulse voltammetry (DPV) quantitative analysis, it is very difficult to measure the peak height of a component in a sample with low concentration. Chen et al. [78] employed a new type of wavelet function known as DOG (the difference of Gaussians) to process DPV signals. In this study, they first transformed the DPV signal of a sample in high concentration to determine a scale parameter, and then transformed the signals of samples with low concentration with the predetermined scale parameter. The results showed that a new linear calibration curve can be obtained and the detection range can be extended in the low concentration side.

Using the edge detection property of WT, an application of WT to determine the endpoint in potentiometric titration was proposed by Wang et al. [79]. In this work, the authors used a second-order differential spline function to process the titration curve, and used the discrete wavelet coefficients to determine the endpoint. Titration curves of HCl, AcOH, H_3PO_4, and $H_2C_2O_4$ were studied by this method. It was found that the endpoint could be determined easily and accurately.

Fourier self-deconvolution is an effective means for resolving overlapping bands, but this method requires a mathematical model to yield the deconvolution and it is quite sensitive to noises in the unresolved bands. A WT-based Fourier deconvolution was proposed by Zhang et al. [80], who obtained a discrete approximation from WT of the original data and substituted it for the original data to be deconvolved and then used another discrete approximation as a lineshape function to yield the deconvolution. After that, they employed the B-spline wavelet, instead of the apodization function, to smooth the deconvolved data to enhance the signal-to-noise ratio (SNR). This method is not adversely affected by noises in the original data as in the Fourier self-deconvolution. The results of this study [80] indicated that resolution can be significantly enhanced, especially for signals with higher noise level. Furthermore, this method does not require a mathematical model to yield the deconvolution.

In order to resolve the overlapping voltammetric peaks that can be described by the $sech^2$ (hyperbolic secant squared) function, a new method known as the *flip shift subtraction method* (FSSM) was proposed by Wu et al. [81]. The method is built on the basis of finding the peak positions using the CWT with the Marr wavelet. To guarantee the accuracy of the determined peak position, a technique known as the *crossed iterative algorithm of continuous wavelet transform and original signal* (CIACWTOS) is proposed to locate the refined peak positions. The calculated results of synthetic peaks and experimental signals both agreed well with the theoretical predictions. In the case of severe noise (SNR = 10), the peak positions can still be obtained under the appropriate dilation parameter. This work demonstrated that CWT is an efficient tool for finding the peak positions using the Marr wavelet even in the case of serious overlap and noise.

The online WT algorithm was adopted in Reference [82] for development of a WT-based voltammetric analyzer. Because the online WT decomposes the sampled signal simultaneously with the progress of sampling, the developed voltammetric analyzer gives all the components contained in the sampled voltammogram. Applications of the equipment in linear sweep voltammetric analysis of mixtures of Pb(II) and Tl(I) and in square-wave voltammetric analysis of mixture of Cd(II) and In(III) were investigated in this study [82]. The results showed that the overlapping peaks of Pb(II) and Tl(I) can be separated easily, and the peak position after the online wavelet transform did not change. The linearity of the calibration curves for Cd(II) and In(III) in the overlapping square-wave voltammograms was retained after on-line WT. Quantitative determination of Cd(II) and In(III) in mixture samples were also investigated, with recovery rates between 92.5% and 107.1%.

5.6.5. Mass Spectrometry

Applications of WT in MS studies are relatively scarce compared with other chemistry fields. WT was applied mainly in two areas: instrumentation design and signal processing of secondary-ion mass spectrometry (SIMS).

With regard to instrumentation design, Shew [83] applied a patent in which WT was applied to process real-time signals from the mass spectrometer. For determination of the relative ion abundances in ion cyclotron resonance mass spectrometry, the author utilized WT to isolate the intensity of a particular ion frequency as a function of position or time within the transient ion cyclotron resonance signal. The WT intensity corresponding to the frequency of each ion species as a function of time can be fitted by an exponential decay curve. Then the fitted curves can be extrapolated back in time to the end of the excitation phase and used to determine accurate values for the relative abundances of the various ions in a sample. By determining the abundances of ions at a point in time at or near the end of excitation, the effects of different rates of decay of the intensity of the signal from different ions species can be reduced and more accurate ion abundance measurements obtained.

Applications of WT in processing SIMS images were studied mainly by researchers from the same group at the Vienna University of Technology [84–91]. In their studies before 1998 [84–87], they focused mainly on the denoising of SIMS iamges. Two-dimensional (2D) WT decomposition was employed in their studies. Different types of wavelet functions, different thresholding methods, and comparison with other conventional denoising methods were investigated. An image compression method was also proposed for three-dimensional (3D) secondary-ion microscopy (SIMS) image sets using a separable nonuniform 3D WT [88]. Compared to different 2D image compression methods, compression ratios of the 3D WT method are about 4 times higher at a comparable peak signal-to-noise ratio (PSNR).

A novel methodology was proposed for automated segmentation of SIMS image sets [89]. The method combines a restoration process (using a combination of channelplate sensitivity compensation with a 3D denoising technique based on the WT) with a fuzzy logic 3D gray-level segmentation that can be used to successfully segment 3D SIMS image sets. The restoration algorithm removes the artifacts produced by the channelplate inhomogeneities as well as noise aberrations from the image sets, and the gray-level thresholding algorithm segments their features.

Image fusion is a process whereby images obtained from various sensors are combined together to provide a more complete picture of the object under investigation. The process of combining SIMS images may be viewed as an attempt to compensate for the inherent effect of SIMS to channel the information obtained from the sample into different images, corresponding to different element phases. WT was proved to be a powerful method for fusion of images by the work in Reference [90], where the use of wavelet-based fusion algorithms on multispectral SIMS images, was discussed, the performance of different wavelet-based fusion rules on different types of image systems was evaluated, and the results were compared to those obtained by conventional fusion techniques.

An "edge" in an image is the boundary between two regions that have relatively distinct properties. Since edges represent the basic structure of an image, detecting edges is of profound importance for image analysis. Wolkenstein et al. [91] discussed the application of a novel edge detection method based on WT for images of chemical content. The results obtained for both simulated and real images proved that the method can detect edges of images even at very low signal-to-noise ratios (SNRs).

These studies are described in more detail in Chapter 20 of Walczak's book [12].

5.6.6. Chemical Physics and Quantum Chemistry

More than 30 papers have been published to reporting studies on applying WT to different chemical physics and quantum chemistry calculations, including quantum mechanics, quantum dynamics, and molecular
mechanics. It was found that there are advantages in using WT in quantum chemistry calculations; for instance, the simultaneous localization of
wavelets in both coordinates and momenta allows one to customize the
basis set to provide resolution locally, the orthonormal wavelets can be
used to eliminate the need to solve a generalized eigenvalue problem, and
the basis sets can be easily extended to multiple dimensions by taking the
tensor products of one-dimensional wavelets.

A detailed summary and discussion of the applications of WT in chemical
physics and quantum chemistry before 1999 can be found in the Chapter
12 of Walczak's book [12], including the applications of WT in calculation of
molecular structure, electromagnetic spectra of chemical molecules, chemical dynamics, chemical kinetics, and fractal structures. In the following
paragraphs, only a few papers published since 1999 are summarized.

A review was published in 1999 by Arias [92]. The paper presented
the theory of wavelets from a physical perspective, provided a unified and
self-contained treatment of nonlinear couplings and physical operators,
and introduced a modern framework for effective single-particle theories of
quantum mechanics. The review focused on electronic structure, and the
advances that are useful for nonlinear problems in the physical sciences in
general were described paper.

Ab initio electronic structure calculations have been proved useful in
understanding nanostructure physics. Extension of the current methods to
large, localized systems remains an active area of research. A key difficulty can be found in the scaling properties associated with the basis sets
used. Therefore, it is worthwhile to consider alternatives to the conventional plane-wave and Gauusian bases. In the work published by Richie
et al. [93], the use of wavelet basis was examined.

An all-electron density-functional (AEDF) program using the "Mexican
hat" wavelet was developed by Han et al. [94]. The AEDF program was
applied to the ab initio all-electron calculations of small molecules as pro-
totype systems, and the construction scheme of multiresolution support
spheres was used to optimize the computational efficiency. H_2, CO, and
H_2O molecules and the $1s$ core-ionized C^*O and CO^* molecules were analyzed in detail. Results showed good agreement obtained with experiments
and other theoretical work.

A numerical procedure based on wavelet collocation was suggested
and developed by Liu et al. [95] for the solution of models for packed-bed

chemical reactors and chromatographic columns. They indicated that the algorithm is stable and convergent and is a simple, accurate, fast, and unified approach. It is not necessary to analyze the model on the basis of particular physical laws to derive a scheme for tracking the steep front. Furthermore, wavelets have the capability of representing solutions at different levels of resolution, which makes them particularly useful for developing hierarchical solutions for chemical processes. The most important advantage of this method is that it simulates the chemical processes with steep fronts more effectively than do conventional numerical methods.

Nagy and Pipek [96] studied multiresolution analysis of density operators, electron density, and energy functionals and proved that, for real physical systems, neither arbitrarily fine nor arbitrarily rough details of the wavefunction and density operators can exist. They also showed that the calculation of both kinetic energy and interaction energy expectation values can be reduced to the determination of some universal functions defined on integer-valued arguments.

Leherte [97] proposed a wavelet-based multiresolution analysis approach to generate low-resolution molecular electron density (ED) distribution functions and compared the performance of the proposed method with that of the other two methods (a crystallography-based formalism and an analytical approach) using the critical point graph representations of the molecular ED distributions for pairwise molecular superpositions. The results showed that, for generation of the ED distribution, the crystallography-based method is the fastest. Wavelet-based multiresolution analysis (WMRA) is the most time-consuming, but it is applicable to any grid representation of a molecular property. The analytical approach offers the advantage of generating functions that are continuously scalable, but requires an analytical description of the molecular property considered. For molecular superpositions, the WMRA approach led to the most consistent superposition results.

5.6.7. Conclusion

Applications of WT in various fields of chemistry have been extensively studied. Besides the contents and the works reviewed in this chapter, there are many other works related to applications of WT in chemistry, such as discriminant analysis, multiscale analysis, multifractal analysis, transition detection, and protein sequences analysis. These studies, have proved that WT based techniques are very efficient tools and have played an important role in chemical signal processing. There is no doubt that an increasing number of chemists will be interested in exploring more application of WT in chemistry and will derive more benefit from using the techniques. WT-based techniques will be one of the most popular methods in the future.

REFERENCES

1. I. Daubechies, "Orthonormal bases of compactly supported wavelets," *Commun. Pure Appl. Math.* **41**:909–996 (1988).

2. S. G. Mallat, "Multifrequency channel decompositions of images and wavelet model," *IEEE Trans. Acoust. Speech Signal Process.* **37**:2091–2110 (1989).

3. S. G. Mallat, "A theory of multiresolution signal decomposition: the wavelet representation," *IEEE Trans. Pattern Anal. Machine Learn.* **11**(9):674–693 (1989).

4. D. Daubechies, *Ten Lectures on Wavelets*, CBMS-NSF Reg. Conf. Ser. Appl. Math., SIAM Press, Philadelphia, 1992.

5. C. K. Chui, *An Introduction to Wavelets*, Academic Press, Boston, 1992.

6. C. K. Chui, *Wavelets: A Tutorial in Theory and Application*, Academic Press, Boston, 1992.

7. C. K. Chui, L. Montefusco, and L. Puccio, *Wavelets: Theory, Algorithms and Applications*, Academic Press, San Diego, 1994.

8. C. K. Chui, *Wavelets: A Mathematical Tool for Signal Processing*, SIAM Press, Philadelphia, 1997.

9. J. C. Hoch and A. S. Stern, *NMR Data Processing*, Wiley, New York, 1996.

10. A. Felinger, *Data Analysis and Signal Processing in Chromatography*, Elsevier Science, Amsterdam, 1998.

11. B. G. M. Vandeginste, D. L. Massart, L. M. C. Buydens, S. de Jong, P. J. Lewi, and J. Smeyers-Verbeke, *Handbook of Chemometrics and Qualimetrics: Part B*, Elsevier Science, Amsterdam, 1998.

12. B. Walczak, *Wavelets in Chemistry*, Elsevier Science, Amsterdam, 2000.

13. M. Bos and E. Hoogendam, "Wavelet transform for the evaluation of peak intensities," *Anal. Chim. Acta* **267**:73–80 (1992).

14. C. R. Mittermayr, B. Lendl, E. Rosenberg, and M. Grasserbauer, "The application of the wavelet power spectrum to detect and estimate $1/f$ noise in the presence of analytical signals," *Anal. Chim. Acta* **388**:303–313 (1999).

15. L. M. Shao, B. Tang, X. G. Shao, G. W. Zhao, and S. T. Liu, "Wavelet transform treatment of noise in high performance liquid chromatography," *Chin. J. Anal. Chem.* **25**:15–18 (1997).

16. C. S. Cai and P. D. Harrington, "Different discrete wavelet transforms applied to denoising analytical data," *J. Chem. Inform. Comput. Sci.* **38**:1161–1170 (1998).

17. X. G. Shao and W. S. Cai, "A novel algorithm of the wavelet packets transform and its application to de-noising of analytical signals," *Anal. Lett.* **32**:743–760 (1999).

18. C. R. Mittermayr, H. Frischenschlager, E. Rosenberg, and M. Grasser-bauer, "Filtering and integration of chromatographic data: A tool to improve calibration?" *Fresenius J. Anal. Chem.* **358**:456–464 (1997).

19. W. S. Cai, Q. Ren, X. G. Shao, Z. X. Pan, and M. S. Zhang, "The compression and de-noising of chromatograms using wavelet neural networks," *Anal. Lab.* **17**(6):1–4 (1998).

20. X. G. Shao, F. Yu, H. B. Kou, W. S. Cai, and Z. X. Pan, "A wavelet-based genetic algorithm for compression and de-noising of chromatograms," *Anal. Lett.* **32**:1899–1915 (1999).

21. Z. X. Pan, X. G. Shao, H. B. Zheng, W. Liu, H. Wang, and M. S. Zhang, "Correction of baseline drift in high-performance liquid chromatography by wavelet transform," *Chin. J. Anal. Chem.* **24**:149–153 (1996).

22. H. L. Shen, J. H. Wang, Y. Z. Liang, K. Pettersson, M. Josefson, J. Gottfries, and F. Lee, "Chemical rank estimation by multiresolution analysis for two-way data in the presence of background," *Chemometr. Intell. Lab. Syst.* **37**: 261–269 (1997).

23. X. G. Shao, W. S. Cai, P. Y. Sun, M. S. Zhang, and G. W. Zhao, "Quantitative determination of the components in overlapping chromatographic peaks using wavelet transform," *Anal. Chem.* **69**:1722–1725 (1997).

24. X. G. Shao, W. S. Cai, and P. Y. Sun, "Determination of the component number in overlapping multicomponent chromatogram using wavelet transform," *Chemometr. Intell. Lab. Syst.* **43**:147–155 (1998).

25. X. G. Shao and W. S. Cai, "An application of the wavelet packets analysis to resolution of multicomponent overlapping analytical signals," *Chem. J. Chin. Univ.* **20**:42–46 (1999).

26. X. G. Shao, S. Q. Hou, N. H. Fang, Y. Z. He, and G. W. Zhao, "Quantitative determination of plant hormones by high-performance liquid chromatography with wavelet transform," *Chin. J. Anal. Chem.* **26**:107–110 (1998).

27. X. G. Shao and L. Sun, "An application of the continuous wavelet transform to resolution of multicomponent overlapping analytical signals," *Anal. Lett.* **34**(2):267–280 (2001).

28. E. R. Collantes, R. Duta, W. J. Welsh, W. L. Zielinski, and J. Brower, "Prepro-cessing of HPLC trace impurity patterns by wavelet packets for pharmaceutical fingerprinting using artificial neural networks," *Anal. Chem.* **69**:1392–1397 (1997).

29. X. G. Shao and S. Q. Hou, "On-line resolution of overlapping chromatograms using the wavelet transform," *Anal. Sci.* **15**:681–684 (1999).

30. X. G. Shao and S. Q. Hou, "An on-line wavelet transform for de-noising of high performance liquid chromatograms," *Anal. Lett.* **32**(12):2507–2520 (1999).

31. L. S. Wang, X. Y. Yang, S. F. Xi, and J. Y. Mo, "Wavelet smoothing and denoising to process capillary electrophoresis signals," *Chem. J. Chin. Univ.* **20**:383–386 (1999).

32. C. Perrin, B. Walczak, and D. L. Massart, "The use of wavelets for signal denoising in capillary electrophoresis," *Anal. Chem.* **73**(20):4903–4917 (2001).

33. B. Schirm, H. Benend, and H. Watzig, "Improvements in pentosan polysulfate sodium quality assurance using fingerprint electropherograms," *Electrophoresis* **22**:1150–1162 (2001).

34. B. K. Alsberg, A. M. Woodward, M. K. Winson, J. Rowland, and D. B. Kell, "Wavelet denoising of infrared spectra," *Analyst* **122**:645–652 (1997).

35. F. T. Chau, J. B. Gao, T. M. Shih, and J. Wang, "Infrared spectral compression procedure using the fast wavelet transform method," *Appl. Spectrosc.* **51**: 649–659 (1997).

36. A. K. M. Leung, F. T. Chau, J. B. Gao, and T. M. Shih, "Application of wavelet transform in infrared spectrometry: Spectral compression and library search," *Chemometr. Intell. Lab. Syst.* **43**:69–88 (1998).

37. W. Liu, J. P. Li, J. H. Xiong, Z. X. Pan, and M. S. Zhang, "The compression of IR spectra by using wavelet neural network," *Chin. Sci. Bull.* (Engl. ed.) **42**(10):822–825 (1997).

38. M. Bos and J. A. M. Vrielink, "The wavelet transform for pre-processing IR spectra in the identification of mono- and di-substituted benzenes," *Chemometr. Intell. Lab. Syst.* **23**:115–122 (1994).

39. P. B. Stark, M. M. Herron, and A. Matteson, "Empirically minimax affine mineralogy estimates from Fourier transform infrared spectrometry using a decimated wavelet basis," *Appl. Spectrosc.* **68**:1820–1829 (1993).

40. B. K. Alsberg, A. M. Woodward, M. K. Winson, J. J. Rowland, and D. B. Kell, "Variable selection in wavelet regression models," *Anal. Chim. Acta,* **368**: 29–44 (1998).

41. J. Trygg and S. Wold, "PLS regression on wavelet compressed NIR spectra," *Chemometr. Intell. Lab. Syst.* **42**:209–220 (1998).

42. U. Depczynski, K. Jetter, K. Molt, and A. Niemoller, "Quantitative analysis of near infrared spectra by wavelet coefficient regression using a genetic algorithm," *Chemometr. Intell. Lab. Syst.* **47**:179–187 (1999).

43. C. R. Mittermayr, H. W. Tan, and S. D. Brown, "Robust calibration with respect to background variation," *Appl. Spectrosc.* **55**:827–833 (2001).

44. B. Walczak, E. Bouveresse, and D. L. Massart, "Standardization of near-infrared spectra in the wavelet domain," *Chemometr. Intell. Lab. Syst.* **36**:41–51 (1997).

45. H. W. Tan and S. D. Brown, "Wavelet hybrid direct standardization of near-infrared multivariate calibrations," *J. Chemometr.* **15**:647–663 (2001).

46. C. V. Greensill, P. J. Wolfs, C. H. Spiegelman, and K. B. Walsh, "Calibration transfer between PDA-based NIR spectrometers in the NIR assessment of melon soluble solids content," *Appl. Spectrosc.* **55**:647–653 (2001).

47. K. S. Park, Y. H. Ko, H. Lee, C. H. Jun, H. Chung, and M. S. Ku, "Near-infrared spectral data transfer using independent standardization samples: A case study on the trans-alkylation process," *Chemometr. Intell. Lab. Syst.* **55**:53–65 (2001).

48. C. L. Stork, D. J. Veltkamp, and B. R. Kowalski, "Detecting and identifying spectral anomalies using wavelet processing," *Appl. Spectrosc.* **52**: 1348–1352 (1998).

49. W. S. Cai, L. Y. Wang, Z. X. Pan, J. Zuo, C. Y. Xu, and X. G. Shao, "Application of the wavelet transform method in quantitative analysis of Raman spectra," *J. Raman Spectrosc.* **32**:207–209 (2001).

50. F. Ehrentreich and L. Summchen, "Spike removal and denoising of Raman spectra by wavelet transform methods," *Anal. Chem.* **73**:4364–4373 (2001).

51. G. Neue, "Simplification of dynamic NMR spectroscopy by wavelet transform," *Solid State Nucl. Magn. Reson.* **5**:305–314 (1996).

52. D. Barache, J. P. Antoine, and J. M. Dereppe, "The continuous wavelet transform, an analysis tool for NMR-spectroscopy," *J. Magn. Reson.* **128**:1–11 (1997).

53. H. Serrai, L. Senhadji, J. D. Decertaines, and J. L. Coatrieux, "Time-domain quantification of amplitude, chemical-shift, apparent relaxation-time T_2^*, and phase by wavelet-transform analysis. Application to biomedical magnetic-resonance spectroscopy," *J. Magn. Reson.* **124**:20–34 (1997).

54. X. G. Shao, H. Gu, W. S. Cai, and Z. X. Pan, "Studies on data compression of 1-D NMR spectra using wavelet transform," *Spectrosc. Spect. Anal.* **19**: 139–141 (1999).

55. X. G. Shao, H. Gu, J. H. Wu, and Y. Y. Shi, "Resolution of the NMR spectrum using wavelet transform," *Appl. Spectrosc.* **54**:731–738 (2000).

56. F. T. Chau, T. M. Shih, J. B. Gao, and C. K. Chan, "Application of the fast wavelet transform method to compress ultraviolet-visible spectra," *Appl. Spectrosc.* **50**:339–349 (1996).

57. H. L. Ho, W. K. Cham, F. T. Chau, and J. Y. Wu, "Application of biorthogonal wavelet transform to the compression of ultraviolet-visible spectra," *Comput. Chem.* **23**:85–96 (1999).

58. J. B. Gao, F. T. Chau, and T. M. Shih, "Wavelet transform method for denoising spectral data from ultraviolet-visible spectrophotometer," *SEA Bull. Math.* **20**:85–90 (1996).

59. W. Liu, J. H. Xiong, H. Wang, Y. M. Wang, Z. X. Pan, and M. S. Zhang, "The recognition of UV spectra by using wavelet neural network," *Chem. J. Chin. Univ.* **18**:860–863 (1997).

60. S. X. Ren and L. Gao, "Simultaneous quantitative analysis of overlapping spectrophotometric signals using wavelet multiresolution analysis and partial least squares," *Talanta* **50**:1163–1173 (2000).

61. J. J. Mao, P. Y. Sun, Z. X. Pan, Q. D. Su, and M. S. Zhang, "Wavelet analysis on photoacoustic spectra of degraded PVC," *Fresenius J. Anal. Chem.* **361**: 140–142 (1998).

62. X. G. Shao, W. Li, G. Chen, and Q. D. Su, "On-line analysis of the photoacoustic spectral signal using wavelet transform," *Fresenius J. Anal. Chem.* **363**:215–218 (1999).

63. X. G. Shao, C. Y. Pang, and Q. D. Su, "A novel method to calculate the approximate derivative photoacoustic spectrum using continuous wavelet transform," *Fresenius J. Anal. Chem.* **367**(6):525–529 (2000).

64. Y. Ding, T. Nanba, and Y. Miura, "Wavelet analysis of X-ray diffraction pattern for glass structures," *Phys. Rev. B* **58**:14279–14287 (1998).

65. X. G. Shao, L. M. Shao, and G. W. Zhao, "Extraction of extended X-ray absorption fine structure information from the experimental data using the wavelet transform," *Anal. Commun.* **35**:135–144 (1998).

66. T. Artursson, A. Hagman, S. Bjork, J. Trygg, S. Wold, and S. P. Jacobsson, "Study of preprocessing methods for the determination of crystalline phases in binary mixtures of drug substances by X-ray powder diffraction and multivariate calibration," *Appl. Spectrosc.* **54**:1222–1230 (2000).

67. D. A. Sadler, D. Littlejohn, P. R. Boulo, and J. S. Soraghan, "Application of wavelet transforms to determine peak-shape parameters for interference detection in graphite-furnace atomic-absorption spectrometry," *Spectrochim. Acta*, Part B, **53B**:1015–1030 (1998).

68. X. G. Ma and Z. X. Zhang, "Correction for background of ICP-AES signal by means of wavelet transform," *Spectrosc. Spect. Anal.* **20**:507–509 (2000).

69. L. Eriksson, J. Trygg, E. Johansson, R. Bro, and S. Wold, "Orthogonal signal correction, wavelet analysis, and multivariate calibration of complicated process fluorescence data," *Anal. Chim. Acta* **420**:181–195 (2000).

70. L. Yan and J. Y. Mo, "Study on new real-time digital wavelet filters to electroanalytical signals," *Chin. Sci. Bull.* **42**(17):1567–1570 (1995).

71. H. Fang and H. Y. Chen, "Wavelet analyses of electroanalytical chemistry responses and an adaptive wavelet filter," *Anal. Chim. Acta* **346**:319–325 (1997).

72. H. Fang, J. J. Xu, and H. Y. Chen, "A new method of extracting weak signal," *Acta Chim. Sin.* **56**:990–993 (1998).

73. H. B. Zhong, J. B. Zheng, Z. X. Pan, M. S. Zhang, and H. Gao, "Investigation on application of wavelet transform in recovering useful information from oscillographic signal," *Chem. J. Chin. Univ.* **19**:547–549 (1998).

74. J. B. Zheng, H. B. Zhong, H. Q. Zhang, D. Y. Yang, Z. X. Pan, M. S. Zhang, and H. Gao, "Application of wavelet transform in retrieval of useful information from $d^2E/dt^2 - t$ Signal," *Chin. J. Anal. Chem.* **26**:25–28. (1998).

75. X. P. Zheng, J. Y. Mo, and P. X. Cai, "Simultaneous application of spline wavelet and Riemann-Liouville transform filtration in electroanalytical chemistry," *Anal. Commun.* **35**:57–59 (1998).

76. X. P. Zheng and J. Y. Mo, "The coupled application of the B-spline wavelet and RLT filtration in staircase voltammetry," *Chemometr. Intell. Lab. Syst.* **45**:157–161 (1999).

77. A. Aballe, M. Bethencourt, F. J. Botana, and M. Marcos, "Using wavelet transform in the analysis of electrochemical noise data," *Electrochim. Acta* **44**:4805–4816 (1999).

78. J. Chen, H. B. Zhong, Z. X. Pan, and M. S. Zhang, "Application of the wavelet transform in differential pulse voltammetric data processing," *Chin. J. Anal. Chem* **24**:1002–1006 (1996).

79. H. Wang, Z. X. Pan, W. Liu, M. S. Zhang, S. Z. Si, and L. P. Wang, "The determination of potentiometric titration end-points by using wavelet transform," *Chem. J. Chin. Univ.* **18**:1286–1290 (1997).

80. X. Q. Zhang, J. B. Zheng, and H. Gao, "Wavelet transform-based Fourier deconvolution for resolving oscillographic signals," *Talanta* **55**:171–178 (2001).

81. S. G. Wu, L. Nie, J. W. Wang, X. Q. Lin, L. Z. Zheng, and L. Rui, "Flip shift subtraction method: A new tool for separating the overlapping voltammetric peaks on the basis of finding the peak positions through the continuous wavelet transform," *J. Electroanal. Chem.* **508**:11–27 (2001).

82. X. G. Shao, C. Y. Pang, S. G. Wu, and X. Q. Lin, "Development of wavelet transform voltammetric analyzer," *Talanta* **50**:1175–1182 (1999).

83. S. L. Shew, *Method and Apparatus for Determining Relative Ion Abundances in Mass Spectrometry Utilizing Wavelet Transforms*, U.S. Patent 5,436,447, (July 25, 1995).

84. H. Hutter, C. Brunner, S. Nikolov, C. Mittermayer, and M. Grasserbauer, "Imaging surface spectroscopy for two- and three-dimensional characterization of materials," *Fresenius J. Anal. Chem.* **355**:585–590 (1996).

85. S. G. Nikolov, H. Hutter, and M. Grasserbauer, "De-nosing of SIMS images via wavelet shrinkage," *Chemometr. Intell. Lab. Syst.* **34**:263–273 (1996).

86. M. Wolkenstein, H. Hutter, and M. Grasserbauer, "Wavelet filtering for analytical Data," *Fresenius J. Anal. Chem.* **358**:165–169 (1997).

87. M. Wolkenstein, H. Hutter, S. G. Nikolov, and M. Grasserbauer, "Improvement of SIMS image classification by means of wavelet denoising," *Fresenius J. Anal. Chem.* **357**:783–788 (1997).

88. M. G. Wolkenstein, and H. Hutter, "Compression of secondary ion microscopy image sets using a three-dimensional wavelet transformation," *Microsc. Microanal.* **6**(1):68–75 (2000).

89. M. Wolkenstein, T. Stubbings, and H. Hutter, "Robust automated three-dimensional segmentation of secondary ion mass spectrometry image sets," *Fresenius J. Anal. Chem.* **365**:63–69 (1999).

90. T. C. Stubbings, S. G. Nikolov, and H. Hutter, "Fusion of 2-D SIMS images using the wavelet transform," *Mikrochimica Acta* **133**:273–278 (2000).

91. M. Wolkenstein, T. Kolber, S. Nikolov, and H. Hutter, "Detection of edges in analytical images using wavelet maxima," *J. Trace Microsc. Tech.* **18**:1–14 (2000).

92. T. A. Arias, "Multiresolution analysis of electronic structure: Semicardinal and wavelet bases," *Rev. Mod. Phys.* **71**:267–311 (1999).

93. D. A. Richie and K. Hess, "Wavelet based electronic structure calculations," *Microelectron. Eng.* **47**(1–4):333–335 (1999).

94. S. W. Han, K. J. Cho, and J. Ihm, "Wavelets in all-electron density-functional calculations," *Phys. Rev. B* **60**:1437–1440 (1999).

95. Y. Liu, I. T. Cameron, and F. Y. Wang, "The wavelet-collocation method for transient problems with steep gradients," *Chem. Eng. Sci.* **55**:1729–1734 (2000).

96. S. Nagy and J. Pipek, "Multiresolution analysis of density operators, electron density, and energy functionals," *Int. J. Quantum Chem.* **84**:523–529 (2001).

97. L. Leherte, "Application of multiresolution analyses to electron density maps of small molecules: Critical point representations for molecular superposition," *J. Math. Chem.* **29**(1):47–83 (2001).

LIST OF WAVELET-RELATED INTERNET WEBSITES

A wavelet tutorial: *http://engineering.rowan.edu/~polikar/wavelets/wttutorial.html.*

A wavelet tutorial: *http://nt.eit.uni-kl.de/wavelet/.*

A practical guide to wavelet analysis: *http://paos.colorado.edu/research/wavelets/.*

A wavelet tutorial from S. Mallat's book: *http://cas.ensmp. fr/~chaplais/Wavetour_ presentation/Wavetour_presentation_US. html.*

Amara's wavelet page—wavelet resources: *http://www.amara.com/current/wavelet. html.*

Wavelets—internet resources: *http://www.cosy.sbg.ac.at/~uhl/wav.html.*

Information on wavelets: *http://lettuce.ms.u-tokyo.ac.jp/mei/wavelet.html.*

WaveLab: *http://www-stat.stanford.edu/~wavelab/.*

The MathWorks—wavelet toolbox: *http://www.mathworks.com/products/wavelet/.*

Wavelet Explorer new-Generation signal and image analysis: *http://www.wolfram. com/products/applications/wavelet/.*

Digital signal processing at Rice University: *http://www-dsp.rice.edu/software/.*

XWPL, the X Wavelet Packet Laboratory: *http://math.yale.edu/pub/wavelets/software/ xwpl/html/xwpl.html.*

WAVEKIT—a wavelet toolbox for MATLAB: *http://www.math.rutgers.edu/~ojanen/ wavekit/.*

Wavelet toolbox: *http://www.comsol.se/products/wavelet/index.php.*

Bibliographies on wavelets: *http://liinwww.ira.uka.de/bibliography/Theory/Wavelets/.*

Wavelet transform in chemistry: *http://fg702-6.abct.polyu.edu.hk/wavelet.html.*

Wavelet papers: *http://www.ee.umanitoba.ca/~ferens/wavelets.html.*

The Wavelet Digest: http://www.wavelet.org/.

APPENDIX

VECTOR AND MATRIX OPERATIONS AND ELEMENTARY MATLAB

A.1. ELEMENTARY KNOWLEDGE IN LINEAR ALGEBRA

A.1.1. Vectors and Matrices in Analytical Chemistry

All the data from spectra, chromatograms, voltammograms, kinetic curves, titration curves, and other sources can be digitized into a series of numbers (see Fig. A.1) that can be represented as a vector in mathematics. Thus, when dealing with a vector in this book, we are working on one-dimensional analytical signals. Further, if several one-dimensional chemical signals are compiled together, then a matrix is formed. Many of the so-called hyphenated instrument technologies, such as high-performance liquid chromatography with a diode array detector (HPLC-DAD), gas chromatography with a mass spectroscopic detector (GC-MS), gas chromatography with an infrared spectroscopic detector (GC-IR), high-performance liquid chromatography with a mass spectroscopic detector (HPLC-MS), and capillary electrophoresis with a diode array detector (CE-DAD) have been introduced to chemical laboratories recently. The data generated by the such instrument can be arranged as a matric in which each row represents a spectrum and each column represents a chromatogram (at a given wavelength, wavenumber, or m/e unit) as illustrated in Figure A.2.

Data obtained by "hyphenated instruments" in chemistry are generally called *two-dimensional* or *two-way data*. They have the following features:

1. They contain information in both chromatograms and spectra.
2. The data matrix for one sample run is usually very large. Sometimes it can be more than 80 megabytes, which is much more difficult to handle;

Chemometrics: From Basics To Wavelet Transform, edited by Foo-tim Chau, Yi-zeng Liang, Junbin Gao, and Xue-guang Shao. Chemical Analysis Series, Vol. 164. ISBN 0-471-20242-8. Copyright © 2004 John Wiley & Sons, Inc.

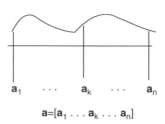

Figure A.1. Illustration of a digitized spectrum.

3. Most important feature of the hyphenated instruments is that they combine a chromatography (e.g. CE, GC, HPLC) and a multichannel spectroscopic detector (e.g. UV, IR, MS). At each regular time interval, a complete spectrum is measured. Consequently, random errors that occur in the chromatogram also influence the spectrum. Apart from these inevitably related fluctuations, the data will also be contaminated by detector noise which could be correlated among adjacent channels. Moreover, noise that is proportional to the magnitude of the signal is more common than purely additive noise. As a result, the overall noise present in the "spectrochromatographic" data collected is complicated by different factors, and more importantly, the noise is heteroscedastic in nature. Thus, the pretreatment of the two-dimensional data is more difficult.

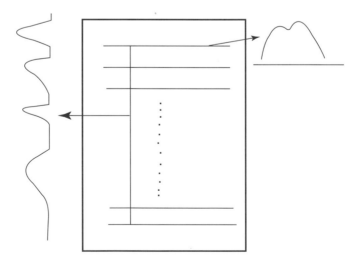

Figure A.2. The physical meanings of the rows and columns of two-dimensional data matrix as generated by a hyphenated instrument.

From this perspective, vectors and matrices in linear algebra are important in mathematical manipulation of one- and two-dimensional data obtained from analytical instruments.

A.1.2.   Column and Row Vectors

A group of real numbers arranged in a column form a column vector, while its transpose is a row vector as shown in the following way:

$$
\mathbf{a} = \begin{bmatrix} a_1 \\ a_2 \\ \vdots \\ a_n \end{bmatrix} \qquad \mathbf{a}^t = [a_1, a_2, \ldots, a_n]
$$

Here, we follow the convention in which a boldfaced variable denotes a column vector or a matrix.

If we say two vectors, $\mathbf{a}$ and $\mathbf{b}$, are equal to each other, that means every corresponding element in them are equal.

A.1.3.   Addition and Subtraction of Vectors

Addition or subtraction of two vectors means that every element of the vectors is added or subtracted in the following manner:

$$
\mathbf{a} \pm \mathbf{b} = \begin{bmatrix} a_1 \pm b_1 \\ a_2 \pm b_2 \\ \vdots \\ a_n \pm b_n \end{bmatrix}
$$

Vector addition and subtraction have the following properties:

$$
\mathbf{a} + \mathbf{b} = \mathbf{b} + \mathbf{a}
$$
$$
(\mathbf{a} + \mathbf{b}) + \mathbf{c} = \mathbf{a} + (\mathbf{b} + \mathbf{c})
$$
$$
\mathbf{a} + \mathbf{0} = \mathbf{a}
$$

Here $\mathbf{0} = [0, 0, \ldots, 0]^t$.

A spectrum of a mixture of two chemical components, say, a and b, can be expressed as the vector sum of the individual spectra $\mathbf{a}$ and $\mathbf{b}$ according to the Lambert–Beer law (see Fig. A.3).

Vector addition of individual spectra to give the spectrum of the mixture can also be applied to other analytical signals such as a chromatogram, a

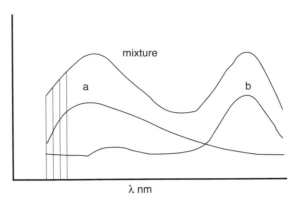

Figure A.3. The mixture spectrum produced by adding two spectra **a** and **b** together.

voltammogram, a kinetic curve, or a titration curve, as they are governed by additive laws similar to the Lambert–Beer law for absorbance. A vector with n elements can be regarded as a point in n-dimensional linear space. Subtraction of two vectors gives the distance between these two points the in the n-dimensional linear space. The geometric meaning of vector subtraction is shown in Figure A.4.

It is well known that addition and subtraction between vectors can be visualized by the so-called parallelogram rule as depicted in Figure A.5.

A.1.4.   Vector Direction and Length

A vector in a n-dimensional linear space has direction and length. The direction of a vector is determined by the ratios between elements. The

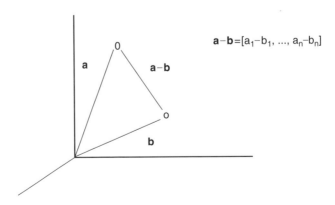

$$\mathbf{a}-\mathbf{b}=[a_1-b_1, ..., a_n-b_n]$$

Figure A.4. Geometric illustration of vector subtraction in an n-dimensional linear space.

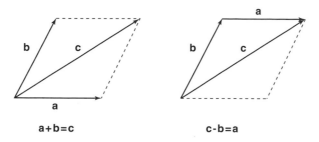

$$a+b=c \qquad\qquad c-b=a$$

Figure A.5. Parallelogram rule for vector addition and subtraction.

length or the magnitudes of a vector is defined by

$$\|a\| = (a_1^2 + \cdots , +a_n^2)^{1/2}$$

In linear algebra, $\|a\|$ is called the *norm* of the vector a.

A.1.5. Scalar Multiplication of Vectors

A vector a multiplied by a scalar (a constant) k is given by

$$ka = \begin{bmatrix} ka_1 \\ ka_2 \\ \vdots \\ ka_n \end{bmatrix}$$

and is called the *scalar multiplication* of a vector in linear algebra. Note that the spectra of different concentrations are just like vectors multiplied by different constants, say, k_1, k_2, and so on (see Fig. A.6).

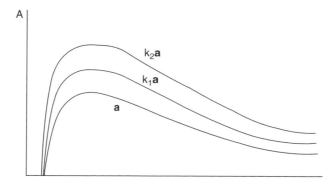

Figure A.6. Profiles obtained by scalar multiplication of a vector (spectrum) by constants k_1 and k_2 with $k_2 > k_1$.

Scalar multiplication of vectors has the following properties:

$$k_1(k_2\mathbf{a}) = (kk_2)\mathbf{a}$$
$$k_1(\mathbf{a} + \mathbf{b}) = k_1\mathbf{a} + k_1\mathbf{b}$$
$$(k_1 + k_2)\mathbf{a} = k\mathbf{a} + k_2\mathbf{a}$$

In particular, we have

$$0\,\mathbf{a} = \mathbf{0} \quad 1\,\mathbf{a} = \mathbf{a} \quad -1\,\mathbf{a} = -\mathbf{a}$$

A.1.6.   Inner and Outer Products between Vectors

When two vectors with the same size (number of elements) multiply each other, there are two possible operations: the inner product and the outer product. The *inner product* (also known as the *dot product* or the *scalar product*) produces a scalar (a number), while the *outer product* (also known as the *cross-product* or the *vector product*) produces a matrix. The following formula (where the superscript *t* denotes transposition) defines the inner product between two vectors:

$$\mathbf{a}^t\mathbf{b} = [a_1, a_2, \ldots, a_n]\begin{bmatrix} b_1 \\ b_2 \\ \vdots \\ b_n \end{bmatrix} = \sum a_i b_i$$

The inner product has the following properties:

$$\mathbf{a}^t(\mathbf{b} + \mathbf{c}) = \mathbf{a}^t\mathbf{b} + \mathbf{a}^t\mathbf{c}$$
$$(\mathbf{a} + \mathbf{b})^t\mathbf{c} = \mathbf{a}^t\mathbf{c} + \mathbf{b}^t\mathbf{c}$$

Figure A.7 gives the geometric meaning of the inner product between two vectors. The inner product is essentially a kind of projection. The

Inner product of vectors

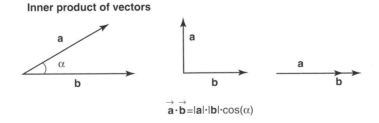

$$\vec{a}\cdot\vec{b} = |a|\cdot|b|\cdot\cos(\alpha)$$

Figure A.7. Graphic representation of inner product of the two vectors **a** and **b**.

concept of projection is very important in chemometrics, and a good understanding of this concept will be very helpful in studying the subject.

If two vectors **a** and **b** are orthogonal with each other, that is, if the angle, α, between them is 90° (as shown in the middle part of Fig. A.7), then the inner product is equal to zero:

$$\mathbf{a}^t\mathbf{b} = 0$$

The outer product of two vectors produces a bilinear matrix of rank equal to 1, which is of special importance in multivariate resolution for two-way data. In the two-way data from "hyphenated" chromatography, every chemical component can be expressed by such a bilinear matrix of rank 1. The outer product of vectors **a** and **b** is given as follows:

$$\mathbf{a}\,\mathbf{b}^t = \begin{bmatrix} a_1 \\ a_2 \\ \vdots \\ a_n \end{bmatrix} [b_1, b_2, \dots, b_n] = \begin{bmatrix} a_1 b_1 & a_1 b_2 & \cdots & a_1 b_n \\ a_2 b_1 & a_2 b_2 & \cdots & a_2 b_n \\ \vdots & \vdots & \cdots & \vdots \\ a_n b_1 & a_n b_2 & \cdots & a_n b_n \end{bmatrix}$$

A.1.7. The Matrix and Its Operations

In general, a matrix is expressed in the following manner

$$\begin{bmatrix} a_{11} & a_{12} & \cdots & a_{1m} \\ a_{21} & a_{22} & \cdots & a_{2m} \\ \vdots & \vdots & \cdots & \vdots \\ a_{n1} & a_{n2} & \cdots & a_{nm} \end{bmatrix}$$

in which there are m columns and n rows.

Usually, capital letters are used to represent matrices, for example, **A**, **B**, ... Lowercase symbols, with integer subscripts i and j, represent the elements in the matrix. For example, a_{ij} in the expression above denotes the matrix r elements at the ith row and the jth column. Thus, sometimes, (a_{ij}) is utilized to denote matrix **A**. Matrix **A** can also be expressed as collection of column vectors:

$$\mathbf{A} = [\mathbf{a}_1, \mathbf{a}_2, \dots, \mathbf{a}_m]$$

A.1.8. Matrix Addition and Subtraction

Two or more matrices of the same order can be added (or subtracted) by adding (or subtracting) their corresponding elements in the following way:

$$\mathbf{A} + \mathbf{B} = (a_{ij}) + (b_{ij}) = (a_{ij} + b_{ij})$$

It is obvious that the addition operation has the following properties:

$$\mathbf{A} + \mathbf{B} = \mathbf{B} + \mathbf{A}$$
$$(\mathbf{A} + \mathbf{B}) + \mathbf{C} = \mathbf{A} + (\mathbf{B} + \mathbf{C})$$

A.1.9. Matrix Multiplication

The product of a matrix of order $(n \times q)$, $\mathbf{A} = (a_{ij})_{n \times q}$ and a matrix $\mathbf{B} = (b_{ij})_{q \times m}$ of order $(q \times m)$ produces a matrix $\mathbf{C} = (c_{ij})_{n \times m}$ of order $(n \times m)$. The elements c_{ij} are defined as

$$c_{ij} = \sum a_{ik} b_{kj}$$

Essentially, c_{ij} is the result of the inner product of the ith row of matrix $\mathbf{A}$ and the jth column of matrix $\mathbf{B}$. It should be noted that matrix multiplication may not satisfy the commutative rule:

$$\mathbf{A}\mathbf{B} \neq \mathbf{B}\mathbf{A}$$

However, it will satisfy the associative rule

$$\mathbf{A}\mathbf{B}\mathbf{C} = (\mathbf{A}\mathbf{B})\mathbf{C} = \mathbf{A}(\mathbf{B}\mathbf{C})$$

and also the distribution rule:

$$\mathbf{A}(\mathbf{B} + \mathbf{C}) = \mathbf{A}\mathbf{B} + \mathbf{A}\mathbf{C}$$
$$(\mathbf{A} + \mathbf{B})(\mathbf{C} + \mathbf{D}) = \mathbf{A}(\mathbf{C} + \mathbf{D}) + \mathbf{B}(\mathbf{C} + \mathbf{D})$$

A.1.10. Zero Matrix and Identity Matrix

In a zero matrix, $\mathbf{0}$, all component elements equal to zero. A square matrix of order $n \times n$ is called an *identity matrix* if all its diagonal elements have unity value and the off-diagonal elements have zero value. It is denoted by $\mathbf{I}$ or $\mathbf{I}_n$ in linear algebra.

It is obvious that the $\mathbf{0}$ and $\mathbf{I}$ matrices have the following features:

$$\mathbf{A} + \mathbf{0} = \mathbf{A}$$
$$\mathbf{I}\mathbf{A} = \mathbf{A}\mathbf{I} = \mathbf{A}$$

A.1.11. Transpose of a Matrix

The transpose of a matrix $\mathbf{A}$, namely, $\mathbf{A}^t$, is obtained by exchanging rows and columns of $\mathbf{A}$:

$$(a_{ij})^t = (a_{ji})$$

From this definition, we have

$$(\mathbf{AB})^t = \mathbf{B}^t \mathbf{A}^t$$
$$(\mathbf{ABC})^t = \mathbf{C}^t \mathbf{B}^t \mathbf{A}^t$$

A matrix is called a *symmetric matrix* if its transpose is equal to itself:

$$\mathbf{A}^t = \mathbf{A}$$

A.1.12. Determinant of a Matrix

The determinant of a square matrix $\mathbf{A}$ of order $(n \times n)$, $|\mathbf{A}|$ or $\det(A)$, is defined by

$$\det(\mathbf{A}) = |\mathbf{A}| = \sum_{i=1}^{n} (-1)^{i+j} a_{ij} |\mathbf{M}_{ij}|$$
$$= \sum_{i=1}^{n} a_{ij} A_{ij} \quad \text{(for any } i, j = \text{a fixed value)}$$

where $|\mathbf{M}_{ij}|$ is the determinant of the minor of the element a_{ij}. The minor $\mathbf{M}_{ij}$ is a $(n-1) \times (n-1)$ matrix obtained by deleting the ith row and the jth column of $\mathbf{A}$. The resulting quantity A_{ij}, is called the cofactor of a_{ij} and is defined as $(-1)^{i+j} |\mathbf{M}_{ij}|$.

Consider the following examples:

$N = 2$:

$$|\mathbf{A}| = a_{11} a_{22} - a_{12} a_{22}$$

$N = 3$: the first column with $j = 1$ is fixed:

$$A_{11} = (-1)^2 \begin{vmatrix} a_{22} & a_{23} \\ a_{32} & a_{33} \end{vmatrix} = (-1)^2 |\mathbf{M}_{11}|$$

$$A_{21} = (-1)^2 \begin{vmatrix} a_{12} & a_{13} \\ a_{32} & a_{33} \end{vmatrix} = (-1)^2 |\mathbf{M}_{21}|$$

$$A_{31} = (-1)^2 \begin{vmatrix} a_{12} & a_{13} \\ a_{22} & a_{23} \end{vmatrix} = (-1)^2 |\mathbf{M}_{31}|$$

and $|\mathbf{A}| = a_{11} A_{11} + a_{21} A_{21} + a_{31} A_{31}$.

As an alternative, one may fix a row and write down the determinant of **A** according to

$$\det(\mathbf{A}) = |\mathbf{A}| = \sum_{j=1}^{n} (-1)^{i+j} a_{ij} |\mathbf{M_{ij}}| \qquad \text{(for any } i, i = \text{a fixed value)}$$

A square matrix **A** is said to be regular or nonsingular if $|\mathbf{A}| \neq 0$. Otherwise **A** is said to be singular.

Let **A** and **B** be $n \times n$ square matrices and k be a scalar; we then have

$$|\mathbf{A}^t| = |\mathbf{A}|$$
$$|k\mathbf{A}| = k^n |\mathbf{A}|$$
$$|\mathbf{AB}| = |\mathbf{A}||\mathbf{B}|$$
$$|\mathbf{A}^2| = |\mathbf{A}|^2$$

If A is a diagonal or triangular matrix, then

$$|\mathbf{A}| = \prod_{i=1}^{n} a_{ii}$$

A.1.13. Inverse of a Matrix

If two square matrices, say, **A** and **B**, satisfy $\mathbf{AB} = \mathbf{I}$, then **B** is called the *inverse matrix* of **A** and is denoted by $\mathbf{A}^{-1}$. If $\mathbf{A}^{-1}$ exists, matrix **A** is a nonsingular matrix or a matrix of full rank. It is easily seen that $\mathbf{A}^{-1}$ exists if and only if **A** is nonsingular.

If the inverses $\mathbf{A}^{-1}$ and $\mathbf{B}^{-1}$ exist, the following expressions hold:

$$(k\mathbf{A})^{-1} = k^{-1}\mathbf{A}^{-1}$$
$$(\mathbf{A}\mathbf{B})^{-1} = \mathbf{B}^{-1}\mathbf{A}^{-1}$$
$$(\mathbf{A}^t)^{-1} = (\mathbf{A}^{-1})^t$$

A.1.14. Orthogonal Matrix

A square matrix **A** is said to be orthogonal if

$$\mathbf{A}^t\mathbf{A} = \mathbf{A}\mathbf{A}^t = \mathbf{I}$$

The orthogonal matrices have the following properties:

$$\mathbf{A}^t = \mathbf{A}^{-1}$$
$$\det(\mathbf{A}) = \pm 1$$

This is because

$$\det(\mathbf{A})\det(\mathbf{A}) = \det(\mathbf{A}^t)\det(\mathbf{A}) = \det(\mathbf{A}^t\mathbf{A}) = \det(\mathbf{I}) = 1$$

A.1.15. Trace of a Square Matrix

The trace of a square matrix, tr($\mathbf{A}$), is defined as the sum of the diagonal elements as

$$\text{tr}(\mathbf{A}) = \sum a_{ii}$$

In a special case when $\mathbf{A}$ is a matrix of order (1×1), it contains only one element a, then

$$\text{tr}(\mathbf{A}) = a$$

For example, a quadratic type $\mathbf{y}^t\mathbf{A}\mathbf{y}$ is a number:

$$\text{tr}(\mathbf{y}^t\mathbf{A}\mathbf{y}) = \mathbf{y}^t\mathbf{A}\mathbf{y}$$

Properties of the trace of a square matrix are as follows:

$$\text{tr}(\mathbf{A} + \mathbf{B}) = \text{tr}(\mathbf{A}) + \text{tr}(\mathbf{B})$$
$$\text{tr}(\alpha\mathbf{A}) = \alpha\,\text{tr}(\mathbf{A})$$
$$\text{tr}(\mathbf{A}\mathbf{B}) = \text{tr}(\mathbf{B}\mathbf{A})$$
$$E[\text{tr}(\mathbf{A})] = \text{tr}[E(\mathbf{A})]$$
$$\text{tr}(\mathbf{A}\mathbf{A}^t) = \text{tr}(\mathbf{A}^t\mathbf{A}) = \sum_{i=1}^{n}\sum_{j=1}^{n} a_{ij}^2$$

It is obvious that if $\mathbf{a} = [a_1, a_2, \ldots, a_n]^t$ is a vector of n elements, then the squared norm may be written as

$$\|\mathbf{a}\|^2 = \mathbf{a}^t\mathbf{a} = \sum_{i=1}^{n} a_i^2 = \text{tr}(\mathbf{a}\mathbf{a}^t)$$

A.1.16. Rank of a Matrix

For matrix $\mathbf{A}$ of order $(n \times m)$, its rank is the number of linearly independent row vectors (or column vectors) in it (see the example below) and is denoted by rank($\mathbf{A}$). It has the following features:

$$\mathbf{A}^t = (\mathbf{A}^{-1})^t$$

$$0 \le \text{rank}(\mathbf{A}) \le \min(n, m)$$

$$\text{rank}(\mathbf{AB}) \le \min[\text{rank}(\mathbf{A}), \text{rank}(\mathbf{B})]$$

$$\text{rank}(\mathbf{A} + \mathbf{B}) \le \text{rank}(\mathbf{A}) + \text{rank}(\mathbf{B})$$

$$\text{rank}(\mathbf{A}^t\mathbf{A}) = \text{rank}(\mathbf{AA}^t) = \text{rank}(\mathbf{A})$$

The rank of a square matrix equals its order n if and only if det($\mathbf{A}$) is not equal to zeros:

$$\text{rank}(\mathbf{A}) = n \qquad (\det(\mathbf{A}) \ne 0)$$

Remarks. When a sample is measured by a "hyphenated instrument," the data can be arranged in the form of a matrix. If there is no measurement noise and the spectrum of every absorbing chemical component is different from all the other spectra, then the rank of the data matrix equals the number of chemical components within the sample.

Example A.1. Suppose that a data matrix is composed of n vectors (spectra) as obtained from measurements that are a linear combination of the vectors $\mathbf{a}$ and $\mathbf{b}$, pure spectra of two chemical components. The rank of this matrix is 2 as there are only two linearly independent vectors in it. Each of the n vectors (spectra) $\mathbf{m}_i$ (with $i = 1, \ldots, n$) can be expressed by the following formula:

$$\mathbf{m}_i = c_{ia}\mathbf{a} + c_{ib}\mathbf{b} \qquad (i = 1, 2, \ldots, n)$$

where c_{ia} and c_{ib} are the relative concentrations of the two components under the ith condition. Thus, the linear space is essentially determined by the two vectors $\mathbf{a}$ and $\mathbf{b}$ of the chemical components as illustrated in Figure A.8.

A.1.17. Eigenvalues and Eigenvectors of a Matrix

For a matrix $\mathbf{A}$, we have the following relationship

$$\mathbf{A}\gamma_i = \lambda_i\gamma_i \qquad (i = 1, 2, \ldots, k)$$

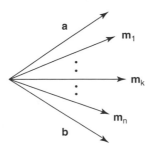

Figure A.8. Geometric illustration of the vectors $\mathbf{m}_1$ and $\mathbf{m}_n$ obtained from the linear combination of two vectors $\mathbf{a}$ and $\mathbf{b}$ (components).

where γ_i ($i = 1, 2, \ldots, k$) are called the *eigenvectors* of the matrix $\mathbf{A}$ while λ_i are the corresponding eigenvalues. If matrix $\mathbf{A}$ is a symmetric matrix, all the eigenvalues are real numbers.

If there is a nonsingular matrix Γ, with $\mathbf{B} = \Gamma\mathbf{A}\Gamma^{-1}$, the square matrices $\mathbf{B}$ and $\mathbf{A}$ are said to be similar. Also, the matrix $\mathbf{B}$ is called the orthogonal similar matrix of $\mathbf{A}$, if Γ is an orthogonal matrix.

Any symmetric matrix $\mathbf{A}$ can be transformed into a diagonal matrix Λ through the orthogonal similar transformation of

$$\Gamma^t\mathbf{A}\Gamma = \Lambda \quad \text{or} \quad \mathbf{A} = \Gamma\Lambda\Gamma^t$$

It can be proved that the diagonal elements of Λ are λ_i ($i = 1, 2, \ldots, k$), the eigenvalues of matrix $\mathbf{A}$, while the column vectors γ_i ($i = 1, 2, \ldots, k$) containing the orthogonal matrix Γ are the corresponding eigenvectors. Moreover, the rank of $\mathbf{A}$ is exactly equal to the number of nonzero real numbers of diagonal elements or eigenvalues of Λ.

A.1.18. Singular-Value Decomposition

For any matrix $\mathbf{A}_{n \times m}$ ($n \geq m$), one can use the technique called *singular-value decomposition* (SVD) to obtain its eigenvalues and eigenvectors. The SVD technique decomposes the matrix into three matrices as $\mathbf{A} = \mathbf{U}\mathbf{S}\mathbf{V}^t$, where $\mathbf{U}$ is the so-called column orthogonal matrix. This means that all columns of $\mathbf{U}$ are orthogonal to each other, that is, $\mathbf{U}^t\mathbf{U} = \mathbf{I}_n$. $\mathbf{S}$ is a diagonal matrix with its diagonal elements equal to the square root of the eigenvalues of covariance matrix of $\mathbf{A}$, while $\mathbf{V}^t$ is a row orthogonal matrix, where the column vectors of $\mathbf{V}$ are orthogonal to one another, or in other words, $\mathbf{V}^t\mathbf{V} = \mathbf{I}_m$. Usually in chemometrics, the matrix $\mathbf{U}$ is called the "scores" while the matrix $\mathbf{V}$ is called the "loadings."

Figure A.9 illustrates the SVD of matrix $\mathbf{A}$.

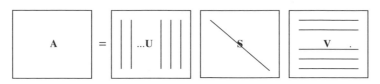

Figure A.9. Illustration of singular value decomposition of a matrix **A**.

Every column in the column orthogonal matrix **U** is the left eigenvector of the matrix **A**, as

$$\mathbf{AV} = \mathbf{US} \quad \Rightarrow \quad \mathbf{A}\mathbf{v}_i = s_i\mathbf{u}_i$$

Every row in the row orthogonal matrix $\mathbf{V}^t$ is the right eigenvector of the matrix **A** because

$$\mathbf{U}^t\mathbf{A} = \mathbf{S}\mathbf{V}^t \quad \Rightarrow \quad \mathbf{u}_i^t\mathbf{A} = s_i\mathbf{v}_i^t$$

A.1.19. Generalized Inverse

For a matrix of order $(n \times m)$, if there is a matrix **B** of order $(m \times n)$, satisfying the equation

$$\mathbf{ABA} = \mathbf{A}$$

then **B** is called a *generalized inverse matrix* of **A** and is denoted by $\mathbf{A}^-$, or sometimes is simply called the "$-$" inverse. It can be easily seen that $\mathbf{A}^-$ is just $\mathbf{A}^{-1}$ if matrix **A** is a full-rank matrix. From this point of view, it can be seen that the generalized inverse is an extension of the inverse which is only defined for square matrices. The question here is whether the generalized inverse is not unique. A unique generalized inverse can be defined by specified constraints, and such a generalized inverse is called the "$+$" inverse or the *Moore–Penrose inverse*, denoted as $\mathbf{A}^+$. It satisfies the following four conditions:

1. $\mathbf{AA}^+\mathbf{A} = \mathbf{A}$
2. $\mathbf{A}^+\mathbf{AA}^+ = \mathbf{A}^+$
3. $(\mathbf{AA}^+)^t = \mathbf{AA}^+$
4. $(\mathbf{A}^+\mathbf{A})^t = \mathbf{A}^+$

The Moore–Penrose inverse has the following features:

$$\text{rank}(\mathbf{A}) = \text{rank}(\mathbf{A}^+)$$
$$(\mathbf{A}^+)^+ = \mathbf{A}$$
$$(\mathbf{A}^t)^+ = (\mathbf{A}^+)^t$$
$$((\mathbf{AB})^+)^t = (\mathbf{A}^t\mathbf{B}^t)^+$$

It should be noted that $\mathbf{AA}^+$ and $\mathbf{A}^+\mathbf{A}$ are both symmetric and idempotent. A matrix is idempotent if the square of the matrix is equal to itself (i.e., $\mathbf{A}^2 = \mathbf{A}$).

A.1.20. Derivative of a Matrix

If the elements in matrix $\mathbf{A}$ are functions of variable t, then the derivative of $\mathbf{A}$ is still a matrix and is denoted as $d\mathbf{A}/dt$. If a_{ij} is the element at the ith row and the jth column, the element in the corresponding derivative matrix $d\mathbf{A}/dt$ is represented by da_{ij}/dt.

The derivative of matrix has the following properties:

$$\frac{d(\mathbf{AB})}{dt} = \left(\frac{d\mathbf{A}}{dt}\right)\mathbf{B} + \left(\frac{d\mathbf{B}}{dt}\right)\mathbf{A}$$
$$\frac{d[\text{tr}(\mathbf{A})]}{dt} = \text{tr}\left(\frac{d\mathbf{A}}{dt}\right)$$

A.1.21. Derivative of a Function with Vector as Variable

Suppose that f is a number function with a vector as its variable, that is, $f = f(\mathbf{x})$; the derivative of f with respect to vector $\mathbf{x}$ is defined as follows:

$$\frac{df(\mathbf{x})}{d\mathbf{x}} = \left[\frac{df}{dx_1}, \frac{df}{dx_2}, \frac{df}{dx_3}, \dots, \frac{df}{dx_n}\right]^t$$

Here $\mathbf{x} = [x_1, x_2, \dots, x_n]^t$. The derivative of a function of vectorial variable has the features mentioned below.

If $\mathbf{a}$ is constant vector, then

$$\frac{d(\mathbf{a}^t\mathbf{x})}{d\mathbf{x}} = \mathbf{a}$$
$$\frac{d(\mathbf{x}^t\mathbf{a})}{d\mathbf{x}} = \mathbf{a}$$
$$\frac{d(\mathbf{x}^t\mathbf{Ax})}{d\mathbf{x}} = 2\mathbf{Ax}$$

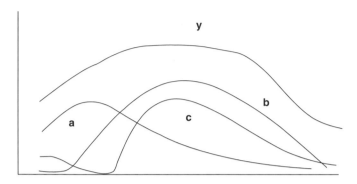

Figure A.10. The spectrum **y** of a mixture that contains three chemial components with their pure spectra **a**, **b**, and **c**.

Here **A** is a symmetric square matrix, and all the elements in matrix **A** are constants.

Example A.2. Suppose that the spectrum of a mixture containing three chemicals, **a**, **b**, and **c** (see Fig. A.10) is measured, and their pure spectra, s_1, s_2, and s_3, are also available. Can we determine the concentrations of these three chemical compounds in the mixture? If "Yes," how can we do it?

According to the Lambert–Beer law, the spectra of the mixture **y** are related to the concentrations of the three components by

$$\mathbf{y} = c_1\mathbf{s}_1 + c_2\mathbf{s}_2 + c_3\mathbf{s}_3 + \mathbf{e}$$

Here c_i ($i = 1, 2, 3$) are the relative concentrations of the components and vector **e** represents the measurement noises, which is usually assumed to be a series of random numbers with normal distribution and zero mean. Since the spectra **y**, s_1, s_2, and s_3 are known, the problem now is to find the concentrations c_i from the preceding equation. It can be solved in the following way.

The equation $\mathbf{y} = c_1\mathbf{s}_1 + c_2\mathbf{s}_2 + c_3\mathbf{s}_3 + \mathbf{e}$ is essentially a combination of linear equations as expressed below:

$$y_1 = c_1 s_{11} + c_2 s_{12} + c_3 s_{13} + e_1$$
$$y_2 = c_1 s_{21} + c_2 s_{22} + c_3 s_{23} + e_2$$
$$\vdots$$
$$y_n = c_1 s_{n1} + c_2 s_{n2} + c_3 s_{n3} + e_n$$

Here $\mathbf{y} = [y_1, y_2, \dots, y_n]^t$, $\mathbf{s}_1 = [s_{11}, s_{21}, \dots, s_{n1}]^t$, $\mathbf{s}_2 = [s_{12}, s_{22}, \dots, s_{n2}]^t$, and $\mathbf{s}_3 = [s_{13}, s_{23}, \dots, s_{n3}]^t$. If we express the linear equations above in matrix form and ignore the noise vector $\mathbf{e}$, then we have

$$\mathbf{y} = \mathbf{Sc}$$

Here vector $\mathbf{c} = [c_1, c_2, c_3]^t$ represents the unknown relative concentrations of the three compounds and matrix $\mathbf{S}$ is a collection of the three spectra:

$$\mathbf{S} = \begin{bmatrix} s_{11} & s_{12} & s_{13} \\ s_{21} & s_{22} & s_{23} \\ \vdots & \vdots & \vdots \\ \vdots & \vdots & \vdots \\ s_{n1} & s_{n2} & s_{n3} \end{bmatrix} = [\mathbf{s}_1 \quad \mathbf{s}_2 \quad \mathbf{s}_3]$$

In order to solve this matrix equation, we need to find the inverse of $\mathbf{S}$. However, matrix $\mathbf{S}$ is not a square matrix, and hence no inverse matrix is available. Now, let us multiply the equation by $\mathbf{S}^t$ in the following manner:

$$\mathbf{S}^t\mathbf{y} = \mathbf{S}^t\mathbf{Sc}$$

Then it is possible to solve this equation with the help of the inverse matrix because matrix $(\mathbf{S}^t\mathbf{S})$ is a square matrix of full rank. In this way, the solution of the equation is expressed in the following form:

$$\mathbf{c} = (\mathbf{S}^t\mathbf{S})^{-1}\mathbf{S}^t\mathbf{y}$$

Remarks

1. Note first that $\mathbf{S}$ is not a square matrix; thus we cannot obtain its inverse matrix directly.
2. We can prove that the solution above is essentially the famous least-squares solution.
3. In Section A.2, the readers will see that the preceding equation can be solved easily using only one statement with the help of MATLAB.

A.2. ELEMENTARY KNOWLEDGE OF MATLAB

The major advantage of MATLAB in signal processing, especially for manipulating two-dimensional signals, is the simplicity with which signals of all types can be generated and visualized. In this section, we will show how to use MATLAB to carry out vector and matrix operations. Some examples will be utilized to illustrate how simple commands can be used to produce meaningful information and plots.

For convenience, in the rest of this appendix, all the matrix and vector quantities will be shown not in the usual boldface type, as above, but in the same typeface (font) as they appear in the MATLAB command window. Readers should encounter no difficulty in determining whether a matrix or vector is used in the text.

In the MATLAB command window, one can directly give MATLAB a command to execute something. For instance, to generate a matrix of order 3×3, just use the following command:

```
>A=[1 2 3;4 5 6;7 8 9]
```

Here the two square brackets denote a matrix with the quantities enclosed, while a space is used between elements within a row and a semicolon is used for separating rows. After pressing the ENTER key, the following matrix will be shown in the MATLAB window:

```
>A =
   1  2  3
   4  5  6
   7  8  9
```

To obtain the inverse of this matrix, simply key in the following command:

```
>B=inv(A);
```

The results will be stored in matrix B, which is the inverse of matrix A. The semicolon at the end of the statement is used to suppress the output appearing in the command window. This is very useful to avoid excessive outputs while running a MATLAB program.

MATLAB script is very powerful and convenient for handling matrix and vector operations. For instance, to solve the least-squares equation $c=(S^tS)^{-1}S^ty$ as discussed in the previous section, just input the matrix S that contains the standard spectra and the response vector y. Then the solution can be easily obtained by the following statement:

```
>c=inv(S'*S)*S'*y
```

The symbols $*$ and $'$ denote multiplication and transpose operation, respectively.

In order to further explain how to use MATLAB to do chemometric calculation, some commonly used commands will be given below.

A.2.1. Matrix Construction

As mentioned before, one can easily construct the matrix by directly inputting the matrix A under the MATLAB command window as follows:

```
>A=[1 3 5 7; 12 3 5 3; 3 5 9 1]
```

Then, MATLAB will store the matrix A:

```
A=
     1   3   5   7
    12   3   5   3
     3   5   9   1
```

You can also type the following statement to produce another matrix a:

```
>a=[2,3,4,6;1,4,5,7]
```

The output matrix a will be as follows:

```
a=
     2   3   4   6
     1   4   5   7
```

Note that every row in matrix is separated by a semicolon.

A.2.2. Matrix Manipulation

A': transpose of matrix A. If matrix A is a complex matrix, its transpose will be the conjugated transformation.

A+B: sum of matrices A and B. This means that the corresponding elements of matrices A and B are summed together. If A and B are scalars, the two numbers are added.

A-B: difference between matrices A and B. This means that the elements in A will be subtracted by the corresponding elements of B.

A*B: multiplication between matrices A and B. A and B can be matrices and/or vectors, if they comply with the rule of matrix multiplication.

A.*B: element-by-element multiplication between matrices A and B, that is, A(i,j)*B(i,j). Note that matrices A and B must be of the same order in this case, unless one of them is a scalar.

A.2.3. Basic Mathematical Functions

MATLAB provides almost all the commonly used mathematical functions.
The only difference between MATLAB and other advanced computer language, such as C, Pascal, FORTRAN, and BASIC (Beginner's All-Purpose
Symbolic Instruction Code), is that variables used in MATLAB are basically
all vectors and matrices. Therefore, the mathematical functions in MATLAB
are operating on the elements of the matrix, for example

```
>A=[123 245 365 ; 345 345 232]
>B=fix(0.45*A)
>C=cos(A)
```

These three commands will give the following results:

```
>A=
      123   245   365
      345   345   232
>B=
       55   110   164
      155   155   104
>C=
     -0.8880   0.9990   0.8391
      0.8391   0.8391   0.8880
```

MATLAB provides the following trigonometric functions and other commonly used functions:

sin: sine function.

cos: cosine function.

tan: tangent function.

asin: arcsine function.

acos: arccosine function.

atan: arctangent function.

sinh: hyperbolic sine function.

cosh: hyperbolic sine function.

tanh: hyperbolic tangent function.

asinh: inverse hyperbolic sine function.

acosh: inverse hyperbolic cosine function.

atanh: inverse hyperbolic tangent function.

abs: absolute value function. abs(X) is the absolute value of the elements of X. When X is complex, abs(X) is the complex modulus (magnitude) of the elements of X.

angle: phase angle function. angle(X) returns the phase angles, in radians, of a matrix with complex elements.

sqrt: square root function. sqrt(X) is the square root of the elements of X. Complex results are produced if X is not positive.

real: complex real part function. real(X) is the real part of X.

round: round toward nearest integer function. round(X) rounds the elements of X to the nearest integers.

fix: round toward zero function. fix(X) rounds the elements of X to the nearest integers toward zero.

floor: round toward minus infinity. floor(X) rounds the elements of X to the nearest integers toward minus infinity;

ceil: round toward plus infinity. ceil(X) rounds the elements of X to the nearest integers toward infinity.

sign: signum function. For each element of X, sign(X) returns 1 if the element is greater than zero, 0 if it equals zero, and −1 if it is less than zero. For complex X, sign(X)=X ./ abs(X).

rem: remainder after division. rem(x,y) is x-y.*fix(x./y) if $y \neq 0$. By convention, REM(x,0) is not a number (NaN in MATLAB). The input x and y must be real arrays of the same size, or real scalars.

gcd: greatest common divisor. G = gcd(A,B) is the greatest common divisor of corresponding elements of A and B. The arrays A and B must contain nonnegative integers and must be the same size (or either can be scalar). gcd(0,0) is 0 by convention; all other GCDs are positive integers.

lcm: least common multiple. lcm(A,B) is the least common multiple of corresponding elements of A and B. The arrays A and B must contain positive integers and must be the same size (or either can be scalar).

exp: exponential function. exp(X) is the exponential of the elements of X, e to the X. For complex Z=X+i*Y, exp(Z)= exp(X)* (cos(Y)+i*sin(Y)).

log: natural logarithm function. log(X) is the natural logarithm of the elements of X. Complex results are produced if X is not positive.

log10: common (base 10) logarithm function. log10(X) is the base 10 logarithm of the elements of X. Complex results are produced if X is not positive.

rat: rational approximation function.   [N,D]=rat(X,tol) returns
    two integer matrices so that N./D is close to X in the sense
    that abs(N./D-X)<= tol*abs(X). The rational approximations are
    generated by truncating continued fraction expansions. Here
    tol=1.e−6*norm(X(:),1) is the default. S=rat(X) or rat(X, tol)
    returns the continued fraction representation as a string. The same
    algorithm, with the default tol, is used internally by MATLAB for format
    rat.

erf: error function. Y=erf(X) is the error function for each element of X.
    X must be real. The error function is defined as

$$erf(x)=2/sqrt(\pi)\int_0^x exp(-t^2)dt.$$

erfinv: inverse error function. X = erfinv(Y) is the inverse error func-
    tion for each element of X. The inverse error functions satisfies
    y=erf(x), for $-1 <= y < 1$ and $-\infty \leq x \leq \infty$.

Note that all the functions mentioned above can be conveniently used
for calculations. This may be one of the most convenient features of the
MATLAB language.

A.2.4. Methods for Generating Vectors and Matrices

The colon operator in MATLAB is one of the most convenient tools for
constructing vectors and/or matrices. For instance, with the following
commands, one can easily obtain a function data table:

>x=[0.0 : 0.2 : 3.0]'; (obtaining the first column in the ans table)
>y=-exp(x).* sin(x); (obtaining the second column in the ans
                                  table)

```
>[x y]ans=
0          0
0.2000   -0.2427
0.4000   -0.5809
0.6000   -1.0288
0.8000   -1.5965
1.0000   -2.2874
1.2000   -3.0945
1.4000   -3.9962
1.6000   -4.9509
```

```
1.8000   -5.8914
2.0000   -6.7188
2.2000   -7.2967
2.4000   -7.4457
2.6000   -6.9406
2.8000   -5.5088
3.0000   -2.8345
```

One can also use the linspace function to produce linearly spaced vectors and the logspace function to generate logarithmically spaced vectors. For instance, `linspace(x1, x2, N)` will gives *N* linearly equally spaced points between `x1` and `x2`.

If we key in the following statement

```
>linspace(.3,5.2,14)
```

the results obtained will be as follows:

```
ans=

  Columns 1 through 7

    0.3000   0.6769   1.0538   1.4308   1.8077   2.1846   2.5615

  Columns 8 through 14

    2.9385   3.3154   3.6923   4.0692   4.4462   4.8231   5.2000
```

MATLAB also provides functions to construct some special matrices. These include

- diag: diagonal matrices and diagonals of a matrix. `diag(V,K)` where V is a vector with *N* components and K is an integer, the functions returns a square matrix of order `N+abs(K)` with the elements of V on the Kth diagonal. `K=0` is the main diagonal, `K > 0` is above the main diagonal; and `K < 0` is below the main diagonal. If X is a matrix, `diag(X)` returns the main diagonal of X. Thus, `diag(diag(X))` returns a diagonal matrix.
- Hadamard: Hadamard matrix. hadamard(N) is a Hadamard matrix of order *N*, that is, a matrix, say, H, with elements 1 or −1 such that $H'*H=N*I_N$ (where I_N denotes the identity matrix of order *N*). A N-by-N ($N \times N$) Hadamard matrix with $N > 2$ exists only if *N* is divisible by 4, that is, `rem(N,4)=0`.

ones: ones array. `ones(N)` returns a matrix of order *N* with all the elements equals to one. `ones(size(A))` gives a matrix of the same size as A with all the elements equal to one.

rand: uniformly distributed random numbers. `rand(N)` is an N-by-N matrix of random numbers that are uniformly distributed in the interval (0.0,1.0). rand with no argument returns a scalar, while `rand(size(A))` is a matrix of the same size as A.

randn: normally distributed random numbers. `randn(N)` is an N-by-N matrix of random numbers that follow the normal distribution with mean zero and unity variance. `randn(size(A))` returns a matrix of the same size as A.

eye: identity matrix. `eye(N)` returns an identity matrix of order *N*. `eye(size(A))` is an identity matrix having the same size as A.

Thus, to produce a random matrix of order 3×5, one can simply key in the following statement:

```
>rand(3,5)
```

The results are as follows:

```
ans=
      0.9501   0.4860   0.4565   0.4447   0.9218
      0.2311   0.8913   0.0185   0.6154   0.7382
      0.6068   0.7621   0.8214   0.7919   0.1763
```

With these functions, we can easily construct data matrices of any size.

A.2.5.   Matrix Subscript System

In order to indicate the position of an element in a matrix, subscripts are always used in mathematics. In principle, MATLAB follows the same rules as those of mathematics. There is no difference between MATLAB and other advanced computer languages. The only difference in MATLAB is that it is possible for MATLAB to use a vector subscript to define submatrix, through which MATLAB makes the matrix operation very convenient. For instance, if A is a matrix of order 10×10, the statement

```
>A(1:5,3)
```

can be used to construct a column vector of order 5×1, which consists of the first five elements in the third column in matrix A. Again, if we key in the statement

```
>X=A(1:5,7:10);
```

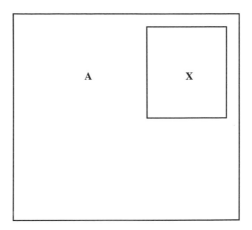

Figure A.11. Generation of a 5 × 4 submatrix using the command X=A(1:5,7:10) to specify the 20 elements (located in an upper right region with label X) within the 10 × 10 matrix **A**.

we can obtain a new matrix of X of order 5 × 4, which contains the elements in the last four columns and in the first five rows in matrix A as shown in Figure A.11, It should be noted that in this expression, if we use only a colon without specifying the starting and ending positions, the command embraces all the rows and/or all the columns of the matrix identified. For example, the statement

```
>A(:,3)
```

gives the third column in matrix A, while the command

```
>A(1:5,:)
```

gives the first five rows in matrix A.

The subscript expression of a matrix can be used in input statements, which makes the matrix operation in MATLAB very convenient. For instance, the following commands can be employed to construct two matrices, B and a:

```
>B=magic(8),
```

```
B =
        64     2     3    61    60     6     7    57
         9    55    54    12    13    51    50    16
        17    47    46    20    21    43    42    24
        40    26    27    37    36    30    31    33
        32    34    35    29    28    38    39    25
        41    23    22    44    45    19    18    48
        49    15    14    52    53    11    10    56
         8    58    59     5     4    62    63     1
```

```
>a=rand(8)

a =
    0.9501  0.8214  0.9355  0.1389  0.4451  0.8381  0.3046  0.3784
    0.2311  0.4447  0.9169  0.2028  0.9318  0.0196  0.1897  0.8600
    0.6068  0.6154  0.4103  0.1987  0.4660  0.6813  0.1934  0.8537
    0.4860  0.7919  0.8936  0.6038  0.4186  0.3795  0.6822  0.5936
    0.8913  0.9218  0.0579  0.2722  0.8462  0.8318  0.3028  0.4966
    0.7621  0.7382  0.3529  0.1988  0.5252  0.5028  0.5417  0.8998
    0.4565  0.1763  0.8132  0.0153  0.2026  0.7095  0.1509  0.8216
    0.0185  0.4057  0.0099  0.7468  0.6721  0.4289  0.6979  0.6449
```

We can replace part of the elements in the matrix a with that of B by the command

```
>a(:,[3 5 7])=B(:,1:3)
```

Then, the matrix a becomes

```
a =
    0.9501  0.8214  64.0000  0.1389   2.0000  0.8381   3.0000  0.3784
    0.2311  0.4447   9.0000  0.2028  55.0000  0.0196  54.0000  0.8600
    0.6068  0.6154  17.0000  0.1987  47.0000  0.6813  46.0000  0.8537
    0.4860  0.7919  40.0000  0.6038  26.0000  0.3795  27.0000  0.5936
    0.8913  0.9218  32.0000  0.2722  34.0000  0.8318  35.0000  0.4966
    0.7621  0.7382  41.0000  0.1988  23.0000  0.5028  22.0000  0.8998
    0.4565  0.1763  49.0000  0.0153  15.0000  0.7095  14.0000  0.8216
    0.0185  0.4057   8.0000  0.7468  58.0000  0.4289  59.0000  0.6449
```

This procedure replaced the third, fifth, and seventh columns of matrix a by the first three columns of matrix B.

In general, if v and w are integer vectors, then A(v,w) represents a submatrix originating from matrix A, in which the rows are determined by vector v, while the columns is determined by vector w. Here vector v is called a *row subscript* and w, the *column subscript*. In this way, some matrix calculations, which may be clumsy to program in other computer languages, can be easily implemented with the help of the subscript system in MATLAB.

Sometimes, we may need to vectorize a matrix before performing some calculations. This can be easily achieved in MATLAB through the following commands:

```
>A=[1 2; 3 4; 5 6]
>b=A(:)
```

The following results can be obtained immediately:

```
A=
   1   2
   3   4
   5   6
b=
   1
   3
   5
   2
   4
   6
```

The reshape function in MATLAB is another way to change the order of a matrix. For example, suppose that we want to change a matrix of order 3×4 into a matrix of order 2×6; this can be achieved by the following commands. First, we define a 3×4 matrix

```
>A=[1 4 7 10; 2 5 8 11; 3 6 9 12]
A=
   1   4   7   10
   2   5   8   11
   3   6   9   12
```

Then, the reshape function is used:

```
>B=reshape(A,2,6)
```

The result is

```
B=
   1   3   5   7    9   11
   2   4   6   8   10   12
```

It is worth noting that MATLAB also defines a special but important matrix, which is the empty matrix. An empty matrix can be constructed by the following statement:

```
>x=[   ]
```

In this way, x is an empty matrix and it can be used as a variable to do the calculation. Using this empty matrix as a variable, one can easily delete some rows and/or columns in a matrix:

```
>A(:,[2,4])=[]
```

The resulting matrix following this operation is that the submatrix of the second column and the fourth column in matrix A is deleted as follows:

```
>A(:,[2,4])=[]
  >A=
        1   7
        2   8
        3   9
```

In MATLAB, some MATLAB functions have their default values for the empty matrix. For instance, functions det (the determinant of a matrix), cond (conditioned number of a matrix), sum (sum of the elements in every column in a matrix), and others have their default values. If X is an empty matrix, then det(X)=1, cond(X)=0 and sum(X)=0. Note that an empty matrix is a very important variable in MATLAB programming.

In order to manipulate matrices easily, MATLAB provides many useful functions. Following are some examples:

max: largest component. For vectors, max(X) is the largest element in X. For matrices, max(X) is a row vector containing the maximum element from each column.

min: smallest component. For vectors, min(X) is the smallest element in X. For matrices, min(X) is a row vector containing the minimum element from each column.

mean: average or mean value. For vectors, mean(X) is the mean value of the elements in X. For matrices, mean(X) is a row vector containing the mean value of each column.

median: median value. For vectors, median(X) is the median value of the elements in X. For matrices, median(X) is a row vector containing the median value of each column.

std: standard deviation. For vectors, std(X) returns the standard deviation. For matrices, std(X) is a row vector containing the standard deviation of each column.

sum: sum of elements. For vectors, sum(X) is the sum of the elements of X. For matrices, sum(X) is a row vector with the sum over each column.

prod: product of elements. For vectors, prod(X) is the product of the elements of X. For matrices, prod(X) is a row vector with the product over each column.

sort: sort in ascending order. For vectors, sort(X) sorts the elements of X in ascending order. For matrices, sort(X) sorts each column of X in ascending order. When X is a cell array of strings, sort(X) sorts the strings in ASCII (American Standard Code for Information Interchange) dictionary order.

Please note that if an N-dimensional array X is passed as the argument, all the functions listed above operate on the first nonsingleton
dimension of X.

conv: convolution and polynomial multiplication. C = conv(A, B)
convolves vectors A and B. The resulting vector is length LENGTH(A)+
LENGTH(B)−1. If A and B are vectors of polynomial coefficients, convolving them is equivalent to multiplying the two polynomials.

corrcoef: correlation coefficients. corrcoef(X) is a matrix of correlation
coefficients formed from array X whose each row is an observation,
and each column is a variable. corrcoef(X,Y), where X and Y are
column vectors, is the same as corrcoef([X Y]).

Consider the following examples. First, let a matrix B be defined and the
outputs of a few functions mentioned above be shown as

```
>B=[1 4 7 10 5; 2 5 8 11 4; 3 6 9 12 6; 2 5 7 3 2]
```

```
>B=
    1     4     7    10     5
    2     5     8    11     4
    3     6     9    12     6
    2     5     7     3     2


>max(B)
ans =
    3     6     9    12     6

>min(B)
ans =
    1     4     7     3     2

>mean(B)
ans =
    2.0000    5.0000    7.7500    9.0000    4.2500

>prod(B)
ans =
    12   600   3528   3960   240
```

```
>sort(B)
ans =
     1     4     7     3     2
     2     5     7    10     4
     2     5     8    11     5
     3     6     9    12     6

>corrcoef(B)
ans =
     1.0000     1.0000     0.8528     0.2000     0.2390
     1.0000     1.0000     0.8528     0.2000     0.2390
     0.8528     0.8528     1.0000     0.6822     0.6625
     0.2000     0.2000     0.6822     1.0000     0.9084
     0.2390     0.2390     0.6625     0.9084     1.0000
```

The most attractive feature of the MATLAB language is that it provides very convenient functions for matrix operations. It makes some programs for processing chemical signals very easy and convenient to implement. Moreover, computations involving MATLAB script are usually very fast and efficient, significantly simplifying and facilitating chemometric programming in the MATLAB language.

A.2.6. Matrix Decomposition

Matrix decomposition is the core of chemometric techniques. Many algorithms used in chemometrics are based on matrix decomposition, such as principal-component analysis (PCA) and partial least squares (PLS). Some familiarity with the basic ideas of matrix decomposition will make it easier to follow the algorithms presented in this book.

A.2.6.1. Singular-Value Decomposition (SVD)

Singular-value decomposition is very important in chemometrics. In MATLAB, matrix decomposition can be performed simply by the following statement

```
>[U,S,V]=svd(A)
```

in which U is a column orthogonal matrix or so-called scores, matrix V is a row orthogonal matrix or so-called loadings, and matrix S is a diagonal matrix satisfying the following relation:

```
A=U*S*V'
```

For example, assume that

```
>A=[1 4 7 10; 2 5 8 11; 3 6 9 12]
```

Then, we have

```
A=
    1  4  7  10
    2  5  8  11
    3  6  9  12
```

Then, we can simply key in

```
>[U,S,V]=svd(A)
```

The following results are obtained:

```
U=
    0.5045    0.7608    0.4082
    0.5745    0.0571   -0.8165
    0.6445   -0.6465    0.4082
S=
   25.4624         0         0  0
         0    1.2907         0  0
         0         0    0.0000  0
V=
    0.1409   -0.8247    0.1605   -0.5237
    0.3439   -0.4263    0.1764    0.8179
    0.5470   -0.0278   -0.8342   -0.0647
    0.7501    0.3706    0.4973   -0.2295
```

A.2.6.2. Eigenvalues and Eigenvectors (eig)

Suppose that A is a square matrix. If a vector x and a scalar a satisfy theequation $Ax=\alpha x$, then x and α are respectively called the *eigenvector* and *eigenvalue* of matrix A. In MATLAB, eigenvalues of A can be obtained by the following statement:

```
>e=eig(A)
```

Here e is a vector containing all the eigenvalues of A. If one also requires the corresponding eigenvectors, then the statement

```
>[V,D]=eig(A)
```

will produce a diagonal matrix D of eigenvalues and a full matrix V whose columns are the corresponding eigenvectors so that $A*V=V*D$. In addition, $E=eig(A,B)$ is a vector containing the generalized eigenvalues of square matrices A and B. The statement $[V,D]=eig(A,B)$ produces a diagonal matrix D of generalized eigenvalues and a full matrix V whose columns are the corresponding eigenvectors so that $A*V=B*V*D$. For example, if

>[x,d]=eig(A'*A)

then, we have

```
    x=
          0.5279   -0.1462   -0.8247    0.1409
         -0.8128   -0.1986   -0.4263    0.3439
          0.0419    0.8356   -0.0278    0.5470
          0.2430   -0.4909    0.3706    0.7501
    d=
         -0.0000         0         0         0
               0    0.0000         0         0
               0         0    1.6658         0
               0         0         0  648.3342
```

A.2.7. Graphic Functions

The MATLAB graphic functions are very powerful and extremely useful in generating scientific plots for data analysis, interpretation, and publication. These functions are very difficult to implement in most of the advanced computer languages without accessing other sophisticated graphic libraries. For instance, mesh plots and contour plots can be easily created in MATLAB. One simple statement can do the job. Here we give a very brief introduction on the graphics features of MATLAB.

We can begin by constructing a matrix of order 16 × 16 and then use this matrix data to show several powerful plotting functions in MATLAB. A small matrix is created first:

```
    >A=  [4   5    3    8
           1   6    9    2
           7   8    2    8
           9   4   12    6]
    A=
           4   5    3   8
           1   6    9   2
           7   8    2   8
           9   4   12   6
```

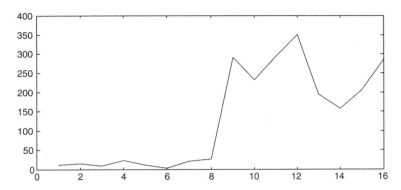

Figure A.12. A plot of the ninth column vector (contains 16 elements) of the matrix **B₁**.

Then, we enlarge the order of the matrix and key in the following statements:

```
>B=[A A';1.2*A sqrt(A)]
>B1=[B 3*B';2.1*B B^2]
```

Finally, we have a matrix of order 16 × 16. If we key in

```
>plot(B1(:,9))
```

then Figure A.12 is generated. If we type the following statements, such as

```
>subplot(221),plot(B1(:,9),'*-')
>hold
```

the command hold retains the current graph so that subsequent plotting commands add to the graph:

```
>subplot(222),plot(B1(:,10),'o-')
>subplot(223),plot(B1(:,11),'+-')
>subplot(224),plot(B1(:,11),'.-')
```

Then, Figure A.13 will appear on the screen.
If we key in the statement

```
>bar(B1(:,6))
```

we will obtain Figure A.14.

To obtain the contour plot of this data matrix, we can simply utilize the following statement.

```
>contour(B1)
```

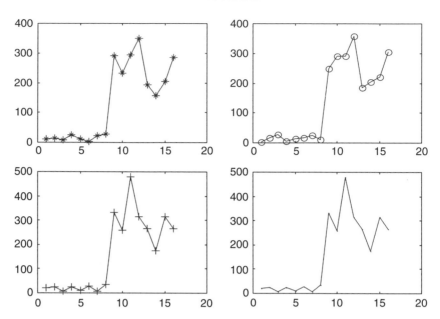

Figure A.13. Plots of the ninth, tenth, and eleventh column vectors of $\mathbf{B}_1$ in the same graph window.

and then, we will have Figure A.15.

To obtain the three-dimensional plot of the data, we can type in

```
>mesh(B1)
```

Then Figure A.16 immediately appears on the screen.

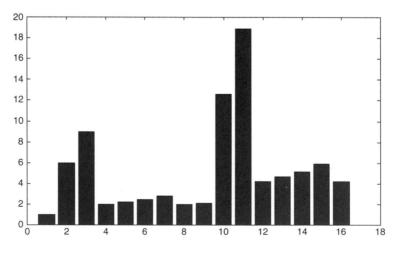

Figure A.14. Bar chart of the sixth column vector of $\mathbf{B}_1$.

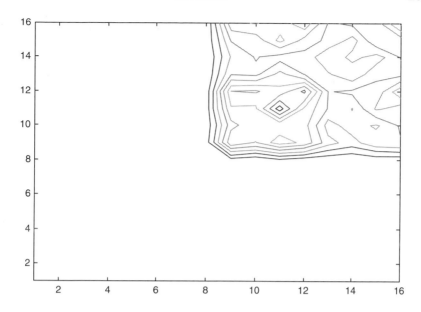

Figure A.15. A contour plot of $\mathbf{B}_1$.

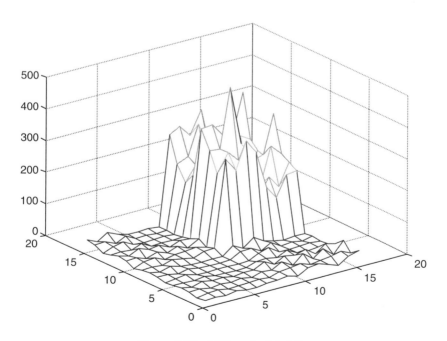

Figure A.16. A 3D mesh plot of $\mathbf{B}_1$.

For these above plots (Figs. A.12–A.16), we can still use functions such as title, `xlabel`, `ylabel`, and axis, to customize the plot such as adding a title, axis range, axis label, or other feature. In summary, the plotting functions in MATLAB are very convenient to use and make visualization of the data and results much easier and simpler.

INDEX

CHEMICAL ANALYSIS

A SERIES OF MONOGRAPHS ON ANALYTICAL CHEMISTRY AND ITS APPLICATIONS

J. D. Winefordner, *Series Editor*